Probability and Mathematical Statistics (Continued)

MUIRHEAD • Aspects of Multivariate Statistical Theory
PURI and SEN • Nonparametric Methods in General Linear Models
PURI and SEN • Nonparametric Methods in Multivariate Analysis
PURI, VILAPLANA, and WERTZ • New Perspectives in Theoretical and Applied Statistics
RANDLES and WOLFE • Introduction to the Theory of Nonparametric Statistics
RAO • Linear Statistical Inference and Its Applications, *Second Edition*
RAO • Real and Stochastic Analysis
RAO and SEDRANSK • W.G. Cochran's Impact on Statistics
RAO • Asymptotic Theory of Statistical Inference
ROBERTSON, WRIGHT and DYKSTRA • Order Restricted Statistical Inference
ROGERS and WILLIAMS • Diffusions, Markov Processes, and Martingales, Volume II: Îto Calculus
ROHATGI • An Introduction to Probability Theory and Mathematical Statistics
ROHATGI • Statistical Inference
ROSS • Stochastic Processes
RUBINSTEIN • Simulation and The Monte Carlo Method
SCHEFFE • The Analysis of Variance
SEBER • Linear Regression Analysis
SEBER • Multivariate Observations
SEN • Sequential Nonparametrics: Invariance Principles and Statistical Inference
SERFLING • Approximation Theorems of Mathematical Statistics
SHORACK and WELLNER • Empirical Processes with Applications to Statistics
STOYANOV • Counterexamples in Probability

Applied Probability and Statistics

ABRAHAM and LEDOLTER • Statistical Methods for Forecasting
AGRESTI • Analysis of Ordinal Categorical Data
AICKIN • Linear Statistical Analysis of Discrete Data
ANDERSON and LOYNES • The Teaching of Practical Statistics
ANDERSON, AUQUIER, HAUCK, OAKES, VANDAELE, and WEISBERG • Statistical Methods for Comparative Studies
ARTHANARI and DODGE • Mathematical Programming in Statistics
ASMUSSEN • Applied Probability and Queues
BAILEY • The Elements of Stochastic Processes with Applications to the Natural Sciences
BAILEY • Mathematics, Statistics and Systems for Health
BARNETT • Interpreting Multivariate Data
BARNETT and LEWIS • Outliers in Statistical Data, *Second Edition*
BARTHOLOMEW • Stochastic Models for Social Processes, *Third Edition*
BARTHOLOMEW and FORBES • Statistical Techniques for Manpower Planning
BECK and ARNOLD • Parameter Estimation in Engineering and Science
BELSLEY, KUH, and WELSCH • Regression Diagnostics: Identifying Influential Data and Sources of Collinearity
BHAT • Elements of Applied Stochastic Processes, *Second Edition*
BLOOMFIELD • Fourier Analysis of Time Series: An Introduction
BOX • R. A. Fisher, The Life of a Scientist
BOX and DRAPER • Empirical Model-Building and Response Surfaces
BOX and DRAPER • Evolutionary Operation: A Statistical Method for Process Improvement
BOX, HUNTER, and HUNTER • Statistics for Experimenters: An Introduction to Design, Data Analysis, and Model Building
BROWN and HOLLANDER • Statistics: A Biomedical Introduction
BUNKE and BUNKE • Statistical Inference in Linear Models, Volume I

(*continued on back*)

Applied Probability and Statistics (Continued)

CHAMBERS • Computational Methods for Data Analysis
CHATTERJEE and HADI • Sensitivity Analysis in Linear Regression
CHATTERJEE and PRICE • Regression Analysis by Example
CHOW • Econometric Analysis by Control Methods
CLARKE and DISNEY • Probability and Random Processes: A First Course with Applications, *Second Edition*
COCHRAN • Sampling Techniques, *Third Edition*
COCHRAN and COX • Experimental Designs, *Second Edition*
CONOVER • Practical Nonparametric Statistics, *Second Edition*
CONOVER and IMAN • Introduction to Modern Business Statistics
CORNELL • Experiments with Mixtures: Designs, Models and The Analysis of Mixture Data
COX • Planning of Experiments
COX • A Handbook of Introductory Statistical Methods
DANIEL • Biostatistics: A Foundation for Analysis in the Health Sciences, *Fourth Edition*
DANIEL • Applications of Statistics to Industrial Experimentation
DANIEL and WOOD • Fitting Equations to Data: Computer Analysis of Multifactor Data, *Second Edition*
DAVID • Order Statistics, *Second Edition*
DAVISON • Multidimensional Scaling
DEGROOT, FIENBERG and KADANE • Statistics and the Law
DEMING • Sample Design in Business Research
DILLON and GOLDSTEIN • Multivariate Analysis: Methods and Applications
DODGE • Analysis of Experiments with Missing Data
DODGE and ROMIG • Sampling Inspection Tables, *Second Edition*
DOWDY and WEARDEN • Statistics for Research
DRAPER and SMITH • Applied Regression Analysis, *Second Edition*
DUNN • Basic Statistics: A Primer for the Biomedical Sciences, *Second Edition*
DUNN and CLARK • Applied Statistics: Analysis of Variance and Regression, *Second Edition*
ELANDT-JOHNSON and JOHNSON • Survival Models and Data Analysis
FLEISS • Statistical Methods for Rates and Proportions, *Second Edition*
FLEISS • The Design and Analysis of Clinical Experiments
FOX • Linear Statistical Models and Related Methods
FRANKEN, KÖNIG, ARNDT, and SCHMIDT • Queues and Point Processes
GALLANT • Nonlinear Statistical Models
GIBBONS, OLKIN, and SOBEL • Selecting and Ordering Populations: A New Statistical Methodology
GNANADESIKAN • Methods for Statistical Data Analysis of Multivariate Observations
GREENBERG and WEBSTER • Advanced Econometrics: A Bridge to the Literature
GROSS and HARRIS • Fundamentals of Queueing Theory, *Second Edition*
GUPTA and PANCHAPAKESAN • Multiple Decision Procedures: Theory and Methodology of Selecting and Ranking Populations
GUTTMAN, WILKS, and HUNTER • Introductory Engineering Statistics, *Third Edition*
HAHN and SHAPIRO • Statistical Models in Engineering
HALD • Statistical Tables and Formulas
HALD • Statistical Theory with Engineering Applications
HAND • Discrimination and Classification
HOAGLIN, MOSTELLER and TUKEY • Exploring Data Tables, Trends and Shapes
HOAGLIN, MOSTELLER, and TUKEY • Understanding Robust and Exploratory Data Analysis
HOCHBERG and TAMHANE • Multiple Comparison Procedures

Component and
Correspondence Analysis

ARS ET SCIENTIA
MUSICA
MELIORA
MEDICINA

Component and Correspondence Analysis

Dimension Reduction by Functional Approximation

Edited by

JAN L.A. VAN RIJCKEVORSEL

Department of Statistics,
TNO Institute for Preventive Health Care
Leiden,
The Netherlands

JAN DE LEEUW

Department of Psychology and Mathematics
UCLA, Los Angeles, USA

JOHN WILEY & SONS

Chichester · New York · Brisbane · Toronto · Singapore

Library of Congress Cataloging in Publication Data:

Component and correspondence analysis: dimension reduction by
functional approximation/edited by Jan L. A. van Rijckevorsel, Jan
de Leeuw.
p. cm. — (Wiley series in probability and mathematical statistics)
Bibliography: p.
Includes index.
ISBN 0 471 91847 4
1. Principal components analysis. 2. Correlation (Statistics)
I. Rijckevorsel, Jan L. A. van II. Leeuw, Jan de. III. Series.
QA278.5.C657 1988
519.5′354—dc19 87–33981
CIP

British Library Cataloguing in Publication Data Available

Typeset by Cotswold Typesetting Ltd., Gloucester
Printed and bound in Great Britain by Anchor Brendon Ltd, Tiptree, Essex

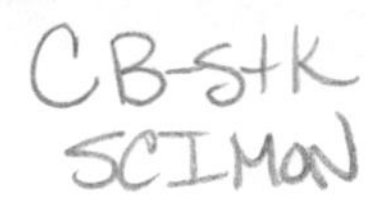

Contents

List of Contributors

Dr Paul Bekker, *Dept of Econometrics, Nettelbosje* 2, *University of Groningen*, 9747 *AE Groningen, The Netherlands.*

Dr Philippe Besse, *Laboratoire de Statistique et Probabilités, Université Paul Sabatier*, 118 *Route de Narbonne*, 31062 *Toulouse Cedex, France*

Dr Jan de Leeuw, *Dept of Psychology and Mathematics, UCLA*, 405 *Hilgard Avenue, Los Angeles, CA* 90024, *USA.*

Dr J.-F. Martin, *Centre de Mathematiques, INSA, bâtiment* 403, 20 *Avenue Albert Einstein*, 69621 *Villeurbanne Cedex, France.*

Dr Jan L.A. van Rijckevorsel, *Dept of Statistics, TNO NIPG, Wassenaarseweg* 56, 2333 *AL Leiden, The Netherlands.*

Dr Suzanne Winsberg, 18 *rue Visconti, Paris* 75006, *France.*

Preface

The technique of correspondence analysis is becoming increasingly popular in many areas of data analysis. In saying this, we have to realize that correspondence analysis is already quite old and that, curiously enough, the popularity of the technique differs greatly from country to country. It has been pointed out by de Leeuw (1983) that Pearson already discovered some of the basic facts connected with correspondence analysis in 1907, while Fisher was the first person who applied correspondence analysis (1940), and who advised others to apply it (Maung, 1941). At about the same time the technique of multiple correspondence analysis was proposed by Guttman (1941). Neither Fisher nor Guttman used the term correspondence analysis, of course, and in fact they did not refer to each other or to earlier work on the same subject. This state of affairs continued when correspondence analysis was rediscovered in the fifties by Burt (1950). Hayashi (1952) introduced the technique in Japan, where it became fairly popular, again under a new name, and with only a few references to the earlier work of Fisher and Guttman. It is not difficult to find at least ten references in the English-speaking applied statistics literature since 1955 in which correspondence analysis is rediscovered, and presented as something completely new. This has led to a great deal of confusion, and a generally quite unsatisfactory state of affairs. The major reason for the amount of disarray in the older literature is perhaps that the technique did not really catch on. Although everybody who studied the problems for which the technique was designed seemed to agree that it provided a very reasonable solution, the computations were simply not feasible until mainframe computing came along in the sixties. Just as factor analysis had its centroid method, whose only justification was its computational simplicity, the scaling of categorical variables had the Likert method, which also could be applied by hand.

The breakthrough came, surprisingly enough, in France. In the mid-sixties Benzécri reinvented the technique, and baptized it 'correspondence analysis'. Because of the peculiar situation in French statistics it rapidly became very popular in France, and in the mid-seventies data analysis and statistics were virtually identical there with correspondence analysis. There are still many French publications in which log-linear analysis or survival analysis is explained by showing how it relates to correspondence analysis. The French language, the highly idiosyncratic style of Benzécri's publications, and the somewhat sectarian and intolerant approach of French Benzécrists to the rest of statistics, did not help very much in spreading the technique to other countries.

This changed in the seventies, for various reasons. Guttman's weighting technique (also known as multiple correspondence analysis) was unearthed by psychometricians such as Lingoes, McDonald, and McKeon. It was now feasible computationally, and it consequently became more interesting. More recently the French approach to correspondence analysis began to attract attention. It has been discussed very completely and competently in the English literature by Greenacre (1984). The books of Lebart, Morineau, and Warwick (1984) and of Jambu (1983) have been translated into English, and there are a large number of papers in statistical journals which compare the French approach to data analysis with the Anglo-American approach (this usually means log-linear analysis). The Japanese contributions were reviewed very ably by Nishisato (1980). By now, classical correspondence analysis and multiple correspondence analysis are quite well known, and anybody who is interested can easily find the relevant literature. A very complete historical review is available in van Rijckevorsel (1987), and the Anglo-French comparisons are summarized and extended in van der Heijden (1987).

But what about optimal scaling? Because these techniques are even more linked with mainframe computing, they are less ancient. In fact one could say that they were hardly thinkable before the computer existed, because they are defined mainly in algorithmic terms. The first optimal scaling techniques were introduced by Shepard and Kruskal in the early sixties. These were actually multidimensional scaling techniques, in which similarity data were related to distances, and least squares loss functions indicating fit were minimized by gradient techniques. It became apparent rather soon that the process of fitting a non-linear model to the data and the process of transforming the data were two components which could be dealt with quite separately from each other. Kruskal, Roskam, and Lingoes wrote families of programs which combined different linear and non-linear models with the idea of optimal scaling (also called optimal quantification or optimal transformation). The separation of the fitting and scaling components of the non-linear techniques was carried out to its logical consequences by de Leeuw, Young, and Takane in the mid-seventies. They used alternating least squares methods to minimize the least squares loss functions, which means that the two algorithmic components were actually completely separate in the program. An example may clarify this. Take a least squares loss function, for example the one used in multiple regression. We choose a particular scaling or transformation of the variables and minimize the function over the model parameters, i.e. the regression coefficients. In the next step we choose transformations from the class of feasible transformations to minimize this very same loss function, keeping the regression coefficients fixed at their current values for the moment. And we alternate these two different minimization steps until convergence. This approach, reviewed by Young (1981), has produced a number of computer programs which can deal with many of the classical multivariate analysis models, and which allow for optimal transformation of each the variables at the same time (where the class of feasible transformation can be different for different variables, i.e. some variables can be transformed monotonically, some polynomially, some arbitrarily, and so on). More recently similar methods have been developed by Breiman and Friedman (1985) and their students, using a particular class of smooth transformations.

The first one to systematically exploit the relationship between correspondence analysis and optimal scaling was de Leeuw (1973). He reviewed almost all of the older literature, and he started a large research program with the explicit purpose of combining these two classes of techniques into a number of useful computer programs, that could be applied profitably in many of the sciences. This resulted in the book by Gifi (1981a, re-issued in 1988) and in a whole series of publications in various books and journals. In this approach multiple correspondence analysis is taken as the starting point to develop a whole class of non-linear multivariate analysis methods, of which the classical methods such as principal component analysis, canonical correlation analysis, multiple regression analysis, discriminant analysis, and path analysis are special (linear) cases. All these special cases are defined by imposing restrictions on the parameters of the multiple correspondence analysis problem. In Gifi the basic idea is to scale the individuals or objects (i.e. to map them into low dimensional Euclidean space) in such a way that individuals with similar response profiles are relatively close together, while individuals with different response profiles are relatively far apart. This is called homogeneity, and it is defined explicitly in terms of Euclidean distances. The emphasis in Gifi (1981a, 1988) is, as in the French literature, on the geometry of the problem, but the ideas are also close to the similarity representations of multidimensional scaling. The restrictions that define principal component analysis as a special case of multiple correspondence analysis, for example, also have a clear geometrical meaning in the space that scales the individuals. This is in contrast with the approach taken by Young (1981) or Breiman and Friedman (1985), who take optimal scaling as their starting point, and emphasize algebraic and analytic properties of solutions.

The authors of the various chapters of this book work in the tradition of the Gifi approach, but they extend the earlier results in many directions. This is most apparent in the first chapters, which are introductory. Bekker and de Leeuw discuss both multiple correspondence analysis (or homogeneity analysis, as it is called by Gifi) and the optimal scaling technique called non-linear principal component analysis. They then proceed to show that in various important special cases these two techniques are closely related, and actually produce the same solutions, although in a quite different form. These special cases are idealized, and not likely to occur exactly in practice, but it turns out that many actual empirical data sets do behave in approximately the same way as these mathematical gauges. This chapter clearly explains the occurrence of the famous horseshoe or Guttman effect in multiple correspondence analysis, and it shows that at least in some applications non-linear principal component analysis is more directly interpretable in terms of the parameters of the original problem.

One of the reasons for going beyond homogeneity analysis is that it is sometimes necessary to analyse continuous variables. In the book by Gifi continuous variables are often discretized in a small number of categories, and this can entail a substantial loss of information. One way of dealing with continuous variables is to use coding schemes that are directly related to classes of feasible transformations defined by splines. This theme is taken up by van Rijckevorsel in Chapter 2. He discusses the various forms of fuzzy coding proposed mainly in the French literature, both in their analytical and geometrical aspects. Splines are a particular form of fuzzy coding.

In the third chapter de Leeuw and van Rijckevorsel take the geometrical approach of Gifi a few steps further, still working in the space in which the individuals are represented, but modifying the definition of homogeneity in various ways. The geometry of using spline functions is explained in this chapter.

In the first French contribution, Besse explains why classical multiple correspondence analysis and principal component analysis are not appropriate for a very common format of continuous data: replicated time series data. Using ideas from functional analysis it is possible to define various alternative metrics which do justice to the fact that the variables are ordered in time. This can be interpreted as yet another variation on the theme of how homogeneity should be defined, and which restrictions are appropriate for which problems. Again splines play an important role in this treatment of dynamic component analysis. Because of the availability of a large variety of transformation functions and/or metrics, the classical question in PCA of how many components should be retained, is modified into which metric, transformation function, filter, . . . etc. leads to the smallest number of well-separated components? Besse illustrates various aspects of this problem.

The last two chapters have a more statistical content. The statistical interpretation of random errors, smoothing or density estimation is one way of incorporating prior information in data analysis. Martin is one of the French statisticians who introduced fuzzy coding. He gives a related probabilistic interpretation, called probability coding or P-coding, in Chapter 5. And analogously to Besse and Winsberg this interpretation of fuzzy coding can also lead to kernel estimation.

In Chapter 6 Winsberg discusses two different approaches to component analysis. First, optimal scaling, which uses the monotone splines introduced by Ramsay and Winsberg into the data analysis literature, is embedded in a more classical statistical framework. This can be seen as bridging another gap: the one between explorative multivariate data analysis and between confirmatory analysis based on statistical modelling, or between confirmatory analysis based on representing the actual data at hand in a pleasing way and the statistical approach of using the likelihood to arrive at inductive generalizations. Besse's procedure for incorporating functional relationships through metrics can also be implemented by using filters on sampled smooth functions. This procedure, originally introduced by Kruskal and Winsberg, is contrasted by Winsberg in the second part of Chapter 6 with the more elaborate approach of Besse and Ramsay.

We do not think, by the way, that it is necessary for us to choose an explicit location on the infamous exploratory–confirmatory continuum. Our point of view is simply that if there is trustworthy prior information available, then it should be incorporated in the data analysis technique, because it will reduce sampling variability of the representation. This prior information can be in the form of a statistical model, but also in the form of restrictions on the correspondence analysis representation. We do not think that the model should always precede the technique. In many situations with which we are familiar there is very little prior information. Imposing restrictions in such situations should give the data analyst a bad conscience, and can lead him rather far astray because he pays for the reduced variability with a large bias, and because he is making confirmatory statements which are at least partly based on figments of his imagination. Choosing between overparametrization and chance capitalization on one side, and between underparametrization and spurious precision on the other side, is a painful process. On the continuum of the various statistical techniques there seems to be an important range where correspondence and component analysis are suitable techniques, which show researchers results in a form in which they want to see them.

Acknowledgements

Since the *initial* planning for this monograph a number of individuals and organizations took part. The essential idea for a book was conceived at a workshop on PCA and splines at the Erasmus University Rotterdam in 1984. The participants were Besse, de Leeuw, Martin and van Rijckevorsel. Valuable assistance in organizing this workshop was given by Dr H. Coolen, and facilities and some funds were provided for by the Erasmus University.

Furthermore The Netherlands Service for Occupational Health and Safety in the Hague offered administrative facilities from 1984–1987 while getting the book organized. Finally the cooperation of the secretarial staff of the Department of Datatheory in typing some of the drafts has been extremely gratifying. We would like to thank especially Marie Verdel for shepherding the manuscript through. Dr P. Besse generously provided his expertise in a particular area, apart from his own contribution. Prof. H. Caussinus smoothed some communication problems.

THE EDITORS

Component and Correspondence Analysis
Edited by J.L.A. van Rijckevorsel and J. de Leeuw

Chapter 1

Relations Between Variants of Non-linear Principal Component Analysis

Paul Bekker
University of Groningen, The Netherlands
and
Jan de Leeuw
UCLA, Los Angeles, USA

INTRODUCTION

Classical principal component analysis can be generalized in different directions to yield non-linear principal component analysis. This means that, in addition to variables measured on an interval scale level, other variables such as ordinal or nominal variables can be analysed in a similar way.

The first generalization is a technique that has many names, but we shall refer to it as 'multiple correspondence analysis'. For an extensive description of multiple correspondence analysis, we refer to Nishisato (1980), Gifi (1981a, 1988), or Greenacre (1984). Another generalization, described by Kruskal and Shepard (1974) and Gifi (1981a, 1988), is 'non-metric principal component analysis'.

In the first section of this chapter a brief account is given of both techniques, which describes them as direct generalizations of principal component analysis. This account is far from being complete and is meant as an introduction only. One can introduce principal component analysis in many different ways. Also, multiple correspondence analysis and non-metric principal component analysis can be looked upon from quite another angle, without regarding them as generalizations of principal component analysis. Both techniques can be presented as multidimensional scaling techniques, using the concept of distance as the central one rather than the concept of correlation.

The principal aim of this chapter is to discuss, in detail, the relationships between multiple correspondence analysis and non-metric principal component analysis. To this end the techniques will be reformulated in terms of optimal scaling in Section 2, and, in the third section, a condition will be derived under which both techniques find essentially the same solution. In Section 4, some important theoretical examples are given for which this condition is satisfied. To conclude, an alternative algorithm for non-linear principal component analysis is applied which combines features of both previous generalizations.

For previous attempts to integrate different generalizations of principal component analysis, we refer to de Leeuw and van Rijckevorsel (1980), Gifi (1981a, 1988), de Leeuw (1982), Tenenhaus and Young (1985). The general approach to PCA in this chapter is taken from Gifi (1981a), the results we discuss are mainly due to de Leeuw (1982).

1. VARIOUS GENERALIZATIONS OF PRINCIPAL COMPONENT ANALYSIS

In this section the concept of homogeneity will serve as a basis for the introduction of multiple correspondence analysis (MCA) and non-metric principal component analysis (NCA) as direct generalizations of ordinary metric principal component analysis (PCA). Homogeneity can be defined on the basis of various models (cf. Fischer, 1974). In this chapter the concept of homogeneity will be used in a data theoretical sense as being closely related to the concept of data reduction (cf. de Leeuw, 1973). In other words, homogeneity deals with the question to what extent different variables measure the same property or properties. In order to answer this question, we need a measure for the difference or resemblance of the variables. On the other hand, the measurement level of the data may allow us to transform the variables before comparing them with each other. By definition, the class of admissible transformations will be different for different types of data. The problem is then to find admissible transformations that maximize the resemblance, or homogeneity, of the variables. When the variables measure more than one property we may want to proceed in order to find another, orthogonal, solution. This is in accordance with the principle of data reduction which advocates that a small number of dimensions should be used to explain a maximum amount of information present in the data.

For the techniques discussed in this chapter, we might say that they share this approach of finding a transformation within a class of admissible transformations such as to minimize the difference between the variables. There are two kinds of distinctions amongst the techniques: either the way the differences between the variables are measured varies, or different classes of admissible transformations are used.

1.1. Homogeneity and linear transformations

Linear transformations of variables may both change their means and their variances. To begin we shall leave the mean out of consideration (we exclude variables which are not in deviations from the mean) and only attach weights to the variables. The difference between the variables will be expressed in a loss function. We want this loss function to attain a minimum when all variables are alike, so one possibility is to use the mean squared Euclidean distance between the transformed variables and one hypothetical common variable (or vector). The resemblance of the transformed variables and the hypothetical variable will be maximized if the loss function is minimized. In that case, the hypothetical variable can serve as a scale for the individuals or objects. The values of this variable are called object scores, component scores or—analogous to factor analysis—factor scores.

The loss function reads:

$$\sigma(x; \phi) = m^{-1} \sum_{j} SSQ(x - \phi_j(h_j)), \tag{1.1}$$

where x is the common variable and ϕ is the transformation of variable $h_j (j=1, \ldots, m)$. In the present case of linear weighting, we can write $\phi_j(h_j) = h_j y_j$, where $y_j (j=1, \ldots, m)$ are the weights for the variables.

We can reformulate the loss function as

$$\sigma(x; y) = x'x + m^{-1}y'Dy - 2m^{-1}x'Hy, \tag{1.2}$$

where H is the data matrix of order $(n \times m)$, D the diagonal of $H'H$ and y the vector of weights. When minimizing this function, we have to impose a restriction on the length of x or y in order to avoid the trivial minimum where both x and y contain zeros.

A possible restriction is $x'x = 1$, which normalizes the object scores. Another restriction where the normalization focuses on the transformed variables is given by $y'Dy = 1$. Both approaches given essentially the same results. Therefore we shall only work out the first normalization.

The minimum of $\sigma(x; y)$, subject to the restriction $x'x = 1$, can be found by minimizing the function $f(x, y, \lambda) = \sigma(x; y) - \lambda(x'x - 1)$, where λ is a Lagrange multiplier. The stationary values can be found in the usual way by equating the partial derivatives to zero.

We find

$$H'x = Dy,$$

$$Hy = m(1 - \lambda)x.$$

Hence $\sigma(x; y) = \lambda$, and the loss function will reach a minimum for the smallest possible value of λ. This λ can be found by combining the equations to

$$HD^{-1}H'x = m(1 - \lambda)x.$$

Clearly, the object scores form a latent vector of $HD^{-1}H'$ and $m(1-\lambda)$ is the latent root. The loss function will thus be minimized if x is the latent vector that corresponds to the largest latent root of $HD^{-1}H'$. If we define the singular value decomposition (SVD-solution, cf. van de Geer, 1986)

$$HD^{-1/2}=V\psi W',$$

where V is of order $(n \times m)$, W is of order $(m \times m)$ and $V'V=I$, $W'W=I$, and ψ is a diagonal matrix of order $(m \times m)$, then we can find the latent roots and vectors in the matrix

$$HD^{-1}H'=V\psi^2 V'.$$

If the latent roots are arranged in descending order of magnitude, the solution is given by $x=v_1$ and $m(1-\lambda)=\psi_1^2$:

$$HD^{-1}H'x=x\psi_1^2. \tag{1.3}$$

At the same time,

$$H'Hy=Dy\psi_1^2. \tag{1.4}$$

Hence,

$$y=D^{-1/2}w_1\psi_1,$$

and

$$y=D^{-1}H'x, \quad \text{or} \quad x\psi_1^2=Hy. \tag{1.5}$$

Which determines completely the solution of the minimization of the loss function.

For variables measuring more than one property, we may want to find another scale x_2 for the individuals, orthogonal to the first scale x_1. This means minimizing $\sigma(x_2; y_2)$ subject to the restriction $x_1'x_2=0$. For a third solution we have to impose $x_1'x_3=0$ and $x_2'x_3=0$, etc.

The successive solutions can now easily be found by the same decomposition as we used for the first solution: $HD^{-1/2}=V\psi W'$, where V is orthogonal, $V'V=I$. We simply use successive singular vectors and values for successive minimizations. The total number of successive solutions equals m, the number of variables. If we collect the vectors x_i and $y_i (i=1, \ldots, m)$ as columns in the matrices X and Y respectively, the simultaneous solution is given by:

$$HD^{-1/2}=XY'D^{1/2}. \tag{1.6}$$

Apparently, the original and transformed (i.e. weighted) variables are linear combinations of the column vectors of X. This means that the vectors $x_i (i=1, \ldots, m)$ form an orthonormal basis of the vector space spanned by the original, or transformed, variables. The sum of the latent roots equals m, for:

$$SSQ(HD^{-1/2})=\mathrm{tr}(HD^{-1}H')=\mathrm{tr}(XY'DYX')$$
$$=\mathrm{tr}(Y'DY)=\mathrm{tr}(\psi^2)=\sum_i \psi_i^2.$$

In the foregoing we did not concern ourselves with the mean of the variables. In order for the transformations to be proper linear transformations, we have to take into account the means. This can be done by decomposing the variables into mean vectors and vectors in deviation from the mean. If we collect these two vectors, as columns, in a matrix F_j for every variable separately, so that $F_j u = h_j$, a linear transformation is given by $F_j z_j$, where z_j is a vector of two weights. The first weight of z_j transforms the mean and the second weight transforms the variable in deviation from the mean. The complete matrix $F=(F_1, \ldots, F_m)$ is of order $(n \times 2m)$ and the vector z is of order $(1 \times 2m)$. Clearly the column vectors of F_j are othogonal, i.e. F'_jF_j is a diagonal matrix. If we now define D to be the diagonal matrix of $F'F$, we can rewrite the loss function (1.1) as:

$$\sigma(x;z)=x'x+m^{-1}\sum_j z'_jF'_jF_jz_j-2m^{-1}\sum_j x'F_jz_j,$$

or

$$\sigma(x;z)=x'x+m^{-1}z'Dz-2m^{-1}x'Fz.$$

Because of the obvious resemblance of this expression to the one in (1.2), the minimization, subject to $x'x=1$, is analogous to the one we already derived for (1.2).

The singular value decomposition $FD^{-1/2}=V\Psi W'$, and the analogous formulations of (1.3), (1.4) and (1.5), will now render $2m$ solutions, since F is of order $(n \times 2m)$. However, if we look more closely at the matrix $FD^{-1/2}$, we see that all columns related to mean vectors are identical to $n^{-1/2}u$ (u being a vector with unit elements only). This totals up to m columns. Consequently we shall find m trivial solutions resulting from transformations of the mean vectors only. It is easy to see that one trivial latent root equals m and $m-1$ trivial latent roots equal 0. For the non-trivial solutions only the variables in deviations from the mean are weighted. Obviously the object scores x_i will be in deviations from the mean too.

Thus, the minimization of the loss function for linear transformations renders m meaningful solutions in deviations from the mean. These solutions correspond to the ones we would have found if we had started with variables in deviations from the mean, and simple weighting would have generated them.

1.2. Principal component analysis: PCA

If, in the foregoing section, we had started with variables in deviations from the mean and normalized to unit length, so that $D=I$ (the identity matrix) and $H'H=R$ (the correlation matrix), then, according to (1.4), (1.5) and (1.6)

$$RY=Y\Psi^2, \quad H=XY', \quad R=YY', \quad X'X=I \quad \text{and} \quad Y'Y=\Psi^2.$$

These formulae relate the latent root and latent vector solutions of the correlation matrix to the principal components X of the data matrix H. In fact, these formulae are well known expressions in principal component analysis (cf. van de Geer, 1986).

1.3. Homogeneity and non-linear transformations: MCA

As we have seen, PCA can be presented as a technique for minimizing differences among variables by transforming these variables linearly. A quite straightforward generalization of PCA in this context is given by extending the class of admissible transformations to include non-linear transformations as well. If we confine ourselves to categorical data, a non-linear transformation is simply found by weighting the categories of a variable. The differences amongst transformed variables can be measured by the same loss function, (1.1), as was used in the foregoing sections to introduce PCA. Minimization of this loss function, for non-linear transformations of discrete variables, adds up to a technique we refer to as multiple correspondence analysis.

If we use so-called indicator matrices, the transformed variables can easily be expressed in matrix notation. An indicator matrix is a binary matrix which indicates the category that an observation is in. Thus, if h_j has k_j categories, the indicator matrix G_j is $n \times k_j$. For the transformation of variable h_j we have,

$$\phi_j(h_j) = G_j y_j,$$

where G_j is the indicator matrix of variable h_j, and y_j is a vector comprising k_j weights for the k_j categories of variable h_j. MCA therefore can be presented as a technique which minimizes the following loss function

$$\sigma(x;\, y)) = m^{-1} \sum_j SSQ(x - G_j y_j), \tag{1.7}$$

or

$$\sigma(x;\, y) = x'x + m^{-1} \sum_j y_j' G_j' G_j y_j - 2m^{-1} \sum_j x' G_j y_j.$$

As the column vectors of G_j are orthogonal, the matrix $D_j = G_j' G_j$ is diagonal and hence we can write

$$\sigma(x;\, y) = x'x + m^{-1} y' D y - 2m^{-1} x' \mathrm{G} y, \tag{1.8}$$

where D is the diagonal supermatrix of univariate marginals and G is the indicator supermatrix. The resemblance with (1.2) is obvious. The minimization is completely analogous: instead of the data matrix H we simply use the indicator supermatrix G.

All solutions will now be rendered by the singular value decomposition of

$$GD^{-1/2} = V \Psi W'.$$

If we write $G'G=C$ for the matrix of bimarginals we have, analogous to (1.3), (1.4) and (1.5)

$$X=V, \quad \text{and} \quad Y=D^{-1/2}W\psi, \tag{1.9}$$

$$GD^{-1}G'X=X\Psi^2, \quad \text{and} \quad CY=DY\Psi^2, \tag{1.10}$$

$$X\Psi^2=GY, \quad \text{and} \quad Y=D^{-1}G'X. \tag{1.11}$$

Since G is of order $(n\times\sum k_j)$, we find $\sum k_j$ solutions. Although some of these, the so-called trivial solutions, are completely meaningless, their existence is, as we saw in Section 1.1, a pleasant circumstance. If we look at the indicator matrices more closely, we see that they comprise a certain amount of redundant information. For, when all vectors of G_j but one, are known, then this one vector is also fixed (we assume that there are no missing data). As a result of these m redundant vectors in G, we shall find in our analysis m trivial solutions. The most prominent trivial solution is given by the vector y_0, where all $\sum k_j$ weights equal $n^{-1/2}$, and the associated object scores in x_0, which also equal $n^{-1/2}$. It is easy to see that this pair x_0, y_0 is a solution; and in fact the loss function reaches its absolute minimum so that there is no loss at all. The corresponding latent root equals m. The $m-1$ remaining trivial solutions are found for weights which are the same for the categories within a variable, but which vary across variables, so that, for all remaining trivial solutions, we have $y'Cy=0$. The corresponding latent roots all equal zero. This situation resembles the situation where we are maximizing homogeneity by means of linear weighting, while the data matrix H comprises units only. As a result of the existence of these trivial solutions, all non-trivial, meaningful solutions are in deviations from the mean. On the one hand we have $X'X=I$, consequently for dimensions we have, $x_o'x_s=0$, or $u'x_s=0$, and so the object scores are in deviations from the mean. On the other hand, we have transformed variables G_jy_j in deviations from the mean.

In order to prove this, we define u_j $(j=1,\ldots,m)$ as vectors comprising k_j units, and $U=u_1 \# \ldots \# u_m$, the direct sum of these vectors. As a result, the m rows of the matrix $U'C$ are all identical to the row vector $u'D$. Consequently we have $U'Cy_s=0$, because $y_0'Dy_s=0$. According to (1.10) we may also write $U'Cy_s=U'Dy_s\psi_s^2$. Then, if $\psi_s^2\neq 0$, this means that $U'Dy_s=0$. So we have for every variable separately $u'D_jy_{js}=0$, or $u'G_j'G_jy_{js}=0$, whence $u'G_jy_{js}=0$, which is a reflection of the fact that the transformed variables for non-trivial solutions are in deviations from the mean. Ultimately we have $\sum k_j-m$ meaningful solutions in deviations from the mean; the inner products of the solutions x_s and G_jy_{js} can now be interpreted in terms of variances, covariances or correlations. For the sum total of the non-trivial latent roots, we find: $\sum\psi_j^2=\sum k_j-m$.

1.3.1. *A relation with* χ^2

We may rewrite the latent vector solutions $D^{-1/2}CD^{-1/2}=W\Psi^2W'$ as a summation of matrices of rank one:

$$D^{-1/2}CD^{-1/2}=\sum_s w_s\psi_s^2w_s'.$$

From this summation ($s=1,\ldots,\sum k_j$) we may remove the trivial matrices for which $\psi_s^2=0$ For the trivial solution for which $\psi_0^2=m$, we have

$$w_0=D^{1/2}y_0\psi_0^{-1}=D^{1/2}u(n\,.\,m)^{-1/2},$$

so that

$$w_0\psi^2w_0'=D^{1/2}uu'D^{1/2}n^{-1}.$$

If we now remove all trivial solutions from W and Ψ, we could write

$$D^{-1/2}(C-Duu'Dn^{-1})D^{-1/2}=W\Psi^2W'. \tag{1.12}$$

For a non-diagonal submatrix of (1.12) we have

$$D_i^{-1/2}(C_{ik}-D_iuu'D_kn^{-1})D_k^{-1/2}. \tag{1.13}$$

C_{ik} being the contingency table of the variables i and k. The matrix $D_iuu'D_kn^{-1}$ contains the expected frequencies on the hypothesis of independence, based on the univariate marginals of C_{ik}. If we multiply (1.13) by the scalar $n^{1/2}$, then its elements equal the difference between the observed and expected frequencies, divided by the root of the expected frequency. This implies that the sum of squares of a non-diagonal submatrix of (1.13) equals $\chi_{ik}^2n^{-1}$ the Pearson χ^2-statistic divided by n.

For diagonal submatrices we have

$$D_i^{-1/2}(D_i-D_iuu'D_in^{-1})D_i^{-1/2}=I-D_i^{1/2}uu'D_i^{1/2}n^{-1}.$$

The sum of squares of this idempotent matrix equals its trace: k_j-1.

The total sum of squares of (1.12) is therefore

$$\sum_i (k_i-i)+\sum\sum_{i\neq k}\chi_{ik}^2n^{-1}=SSQ(W\Psi^2W')=\sum_j \psi_j^4.$$

Since $\sum_j \psi_j^2=\sum_j k_j-m$, we may now write

$$\sum\sum_{i\neq k}\chi_{ik}^2=n\sum_j(\psi_j^4-\psi_j^2).$$

As $\sum_j 1=\sum_j k_j-m$, we also have

$$\sum\sum_{i\neq k}\chi_{ik}^2=n\sum_j(\psi_j^4-2\psi_j^2+1)=n\sum_j(\psi_j^2-1)^2. \tag{1.14}$$

In case of independently distributed variables, the statistic

$$\sum_{i<k} \chi_{ik}^2 = \tfrac{1}{2} n \sum_j (\psi_j^2 - 1),$$

converges to a χ^2-distribution with $\mathrm{df} = \tfrac{1}{2}\{\left(\sum_j k_j - m\right)^2 - \sum_j (k_j - 1)^2\}$ (cf. de Leeuw, 1973).

1.3.2. *The geometry of MCA*

As was the case with PCA, the column vectors of X form an orthonormal basis of a vector space, in which all original variables, the column vectors G, and transformed variables, $G_j y_{js}$, are contained. For,

$$G = XY'D, \quad \text{and} \quad G_j y_{js} = XY_j' D_j y_{js}, \tag{1.15}$$

where Y_j is a matrix of order $(k_j \times \sum k_j)$, with y_{js} $(s = 1, \ldots, \sum k_j)$ as columns. Since the transformed variables $G_j y_{js}$ are in deviations from the mean for non-trivial solutions, the trivial vectors of X and Y_j can be removed without causing any trouble. The transformed variables can thus be represented by vectors in a vector space, of which the x_s $(S = 1, \ldots, (\sum k_j - m))$ form an orthonormal basis.

The squared norms of the transformed variables are usually called discrimination measures: $y_{js}' D_j y_{js}$; the norm of a transformed variable $G_j y_{js}$ is larger as the discrimination between the categories, according to the quantifications y_{js}, is larger.

The sum total of the discrimination measures of all variables, for one solution s, equals $y_s' D y_s = \psi_s^2$. So, for every solution s, the sum total of the discrimination measures is maximized. At the same time the discrimination measure equals the squared correlation between x_s, and $G_j y_{js}$. Namely,

$$y_s = D^{-1} G' x_s, \quad \text{and} \quad y_{js} = D_j^{-1} G_j' x_s, \tag{1.16}$$

from which we may infer that the projection of x_s on the subspace spanned by the column vectors of G_j, is identical to the transformed variable:

$$G_j D_j^{-1} G_j' x_s = G_j y_{js}.$$

As x_s is normalized such that $x_s' x_s = 1$, we have the result that the squared correlation between x_s and $G_j y_{js}$ equals the squared norm of $G_j y_{js}$, which is the same as the discrimination measure.

This derivation indicates that we could interpret MCA as follows. Find, in the space spanned by the column vectors of G, a vector x_1 for which the sum of the squared norms of the projections on the m subspaces G_j is maximized. Having found such a vector x, find another one, subject to the restriction $x_2' x_1 = 0$, etc. The projections of the trivial solution x_0 on the various subspaces G_j all equal x_0

itself; x_0 is contained in the intersection of the subspaces spanned by $G_j (j=1, \ldots, m)$.

Another description is the following. As we have $y_s = D^{-1}G'x_s$, this means that the object scores, corresponding to a certain category of a variable, have a centre of gravity that coincides with the quantification of that category; these points are usually called barycentra. If we replace the object scores by the barycentra of a given variable, the dispersion of the points will be smaller than before replacement. This reduction in dispersion is due to the fact that the dispersion of the object scores around their barycentra has not been taken into account. In this context, the discrimination measure of a variable gives the percentage of dispersion explained by the barycentra. So, for the first dimension, the dispersion of object scores around their centres of gravity is minimized for all variables simultaneously. The second dimension is the best orthogonal dimension, etc.

We can also conceive of the category scores as vectors. For we can write $GD^{-1} = XY'$, and the projections of these vectors on the space spanned by x_1 and x_2, can be completely represented by the category quantifications of these dimensions. Bearing in mind that all column vectors of $GD^{-1/2}$ have norms equal to unity, it is evident that, considering the norms of the column vectors of GD^{-1}, categories with low frequencies are represented by vectors with large norms. Since the projections on the trivial dimension x_0, all have norms equal to one: $u'GD^{-1} = u'DD^{-1} = u'$, these differences in norms will be present in the nontrivial space as well. As there is no reason why those differences in norms should not be present in the first two or more dimensions, we generally expect categories with low frequencies to have extreme positions, when represented as points in two (or more) dimensional space.

Because $GD^{-1} = XY'$, the norms of the projections of the object scores, the rows of X, on the category scores, the rows of Y, have to be proportional to the elements in GD^{-1}, or G. Since the mean is not mapped into non-trivial space, we expect objects corresponding to a certain category of a variable to have a position in the direction of the related category, while others have a position in the opposite direction. Obviously this will be approximately true for two dimensions. But, as we have seen, category scores are in the centre of gravity of object scores and this relates to another possible introduction, or interpretation, of MCA.

1.3.3. *The method of reciprocal averages*

This method begins with the notion that the category quantifications and the object scores in a way should be proportional to one another. For example, the objects should be located in the centre of the categories to which they correspond. This means that $x = Gy/m$. Or, conversely, the situation we already had at hand, the categories should be located in the centre of the objects: $y = D^{-1}G'x$. For non-trivial solutions these two requirements are inconsistent, and so we only require the following proportionalities:

$$x \div Gy/m, \quad \text{and} \quad y \div D^{-1}G'x. \tag{1.17}$$

Consequently we must have,

$$x \div GD^{-1}G'x.$$

From which it is evident that x should be a latent vector of $GD^{-1}G'$. This means that we are dealing with MCA.

Conditional to the normalizations $x'x=1$, or $y'Dy=1$, we find $y=D^{-1}G'x$, and $x\psi^2/m=Gy/m$, or, respectively, $x=Gy/m$, and $y\psi^2/m=D^{-1}G'x$.

1.3.4. *PCA revisited*

As we have seen MCA and PCA are closely related. Beginning with a similar loss function, the presentation of both techniques can be analogous. Partly based on this analogy we can relate PCA and MCA in another way. Every MCA solution generates m transformed variables $G_jy_{js}(j=1, \ldots, m)$. It must therefore be possible to apply PCA to the correlation matrix of these transformed variables. In addition we can apply PCA to every non-trivial MCA solution. In doing so we obtain $\sum k_j \quad m$ correlation matrices, each with m PCA solutions. Thus we find no less than $m\ (\sum k_j - m)$ different solutions. Albert Gifi (1981a) refers to this phenomenon as 'data production' as opposed to 'data reduction'

It would make sense to investigate whether or not redundant information is present in these correlation matrices. Although this topic will be discussed extensively in the following sections, we can already give a relation between the MCA and related PCA solutions.

Using the loss function (1.1) it is easy to see that the first non-trivial MCA solution (i.e. the solution having the largest latent root) equals the first PCA solutions of the associated correlation matrix. The transformed variables of the first MCA solutions, $G_jy_{j1}\ (j=1, \ldots, m)$, all are in deviations from the means and have norms equal to the root of their discrimination measures, $v_{j1}^{1/2}$ say. The loss function

$$\sigma(x_1; y_1) = m^{-1} \sum_j SSQ(x_1 - G_jy_{j1}),$$

has a non-trivial minimum. PCA applied to these transformed variables means minimizing the loss function,

$$\sigma(x; a) = m^{-1} \sum_j SSQ(x - G_jy_{j1}v_{j1}^{-1/2}a_j). \tag{1.18}$$

Clearly then, a minimum will be attained for $x=x_1$ and $a_j=v_{j1}^{1/2}$, and we have a solution identical to the one found for MCA. We may say that the first MCA-solution is found for a transformation which maximizes the largest latent root of the associated correlation matrix. Also the discrimination measures of the first MCA-solution are in fact, using PCA terminology, the squared component loadings of the first component. Of course, the same unambiguous relation holds for the MCA solution with the smallest latent root and the smallest latent root of the corresponding correlation matrix.

For the intermediate MCA solutions things are more complicated. Subsequent MCA solutions are found subject to the restrictions of orthogonality. Applying PCA to the correlation matrix of transformed variables then means substituting the transformations and discrimination measures into (1.18), and any x that minimizes (1.18) will do, i.e. restrictions of orthogonality no longer exist.

Nevertheless, the intermediate MCA solutions do correspond to a principal component of the associated correlation matrix. This component, however, does not necessarily correspond to the largest or smallest latent root. In order to prove this we let v_s denote the vector of discrimination measures and let V_s denote the diagonal matrix comprising the discrimination measures in its diagonal, and $Q_s = \{G_j y_{js}\}$ the matrix of transformed variables. From Section 1.3.2, we know that the correlation between transformed variables and object scores equals the root of the discrimination measures. Then clearly

$$V_s^{-1/2} Q_s' x_s = v_s^{1/2}.$$

Hence,

$$\begin{aligned} R_s v_s^{1/2} &= R_s V_s^{1/2} u = V_s^{-1/2} Q_s' Q_s V_s^{-1/2} V_s^{1/2} u = V_s^{-1/2} Q_s' G y_s \\ &= V_s^{-1/2} Q_s' x_s \psi_s^2 = v_s^{1/2} \psi_s^2. \end{aligned}$$

Obviously $v_s^{1/2}$ is a latent vector of the corresponding correlation matrix; and so $a_j = v_{js}^{1/2}$ generates one of the successive solutions of (1.18).

1.4. Non-metric principal component analysis: NCA

As we have shown, MCA can be regarded as a direct generalization of PCA. Both techniques can be described in terms of a similar loss function, while MCA has an extended class of admissible transformations.

NCA will now be presented as a technique for which the same class of admissible transformations is used as was used for MCA, but which differs from the latter because it starts off with a different loss function.

Nevertheless NCA can still be regarded as a direct generalization of PCA because both loss functions produce the same results in case of linear weighting.

1.4.1. *PCA based on another loss function*

PCA was introduced by using loss function (1.1). The difference among the transformed variables was measured by the mean squared distance to one hypothetical variable. Of course, we could as well base our loss function on the distance of the transformed variables to a plane, or more generally, to a p-dimensional vector space. By doing so, the objects or individuals can be characterized by object scores in more dimensions, although the variables are transformed only once. We can depict the p-dimensional space by a matrix of basis vectors X of order $(n \times p)$, with $X'X = I$. A vector in this space is given by Xa, where a is a vector of p weights, or coordinates. A possible loss function is

$$\sigma(X, a, \phi) = m^{-1} \sum_j SSQ(Xa_j - \phi_j(h_j)). \tag{1.19}$$

For PCA we have $\phi_j(h_j) = h_j y_j$. In order to avoid meaningless solutions, we have to normalize again. The normalization $X'X = I$ is not enough, for both a_j and y_j are not normalized. A possible normalization is to have $y_j' h_j' h_j y_j = 1$; this normalization determines the vectors $h_j y_j$ completely in the matrix $HD^{-1/2}$. If we collect the vectors a_j as rows in the $(m \times p)$ matrix A, we could rewrite (1.19) as

$$\sigma(X, A) = m^{-1} SSQ(HD^{-1/2} - XA'). \tag{1.20}$$

How to minimize this function is known from the Eckart and Young (1936) theorem. If we define the singular value decomposition $HD^{-1/2} = V \Psi W'$, the loss function is minimal for

$$XA' = \sum_i v_i \psi_i w_i'. \tag{1.21}$$

This simultaneous solution for more dimensions is similar to the successive solution described before. This means that the simultaneous solutions for different p are nested; that is to say that the solution for $p = k$ corresponds to the first k principal components found for $p \geq k$.

1.4.2. *Principal components and non-linear weighting*

For NCA the transformed variables are given by $G_j y_j$, just as they were for MCA. Now the loss function (1.19) reads

$$\sigma(X, a, y) = m^{-1} \sum_j SSQ(Xa_j - G_j y_j). \tag{1.22}$$

Again we use the normalizations $X'X = I$ and $y_j' D_j y_j = 1$, the transformed variables are normalized to unit length. However, this does not determine the transformed variables $G_j y_j$. The $G_j y_j$ are to be found so as to maximize the sum of the p largest latent roots of the corresponding correlation matrix. For PCA, or linear weighting, we found the same correlation matrix for every 'transformation'. Now they are different. This also means that the correlation matrix found for $p = k$ will usually be different from the one found for $p \neq k$. Consequently, the solutions are no longer nested. We may observe that NCA for $p = 1$, corresponds to the first MCA solution. However, in general it is not true that NCA with $p = k$ corresponds to the first k MCA solutions. We shall come back to this point in the following sections.

For the minimization of (1.22) we have to use an iterative procedure. For example:

(1) Take an arbitrary vector y for which $y_j' D_j y_j = 1$, and $u' G_j y_j = 0$.

(2) For fixed Gy we can apply principal component analysis of which only the first p solutions will be used.
(3) Subsequently the vectors Xa_j are projected onto the subspaces G_j. The projections are $G_j z_j = G_j D_j^{-1} G_j' X a_j$. Every vector in G_j will be orthogonal to the vector $G_j z_j - Xa_j$, and so must be $G_j y_j - G_j z_j$. Since $G_j y_j - Xa_j = (G_j z_j - Xa_j) + (G_j y_j - G_j z_j)$, we have

$$SSQ(G_j y_j - Xa_j) = SSQ(G_j z_j - Xa_j) + (y_j - z_j)' D_j (y_j - z_j).$$

The first term on the right-hand side is fixed and so we only have to minimize the second term. This can be done simply by setting y equal to z. In the case of ordinal data we can carry out a monotonic regression.
(4) Return to the second iteration step or stop.

Loss function (1.22) is used by Kruskal and Shepard (1974), Tenenhaus (1977) and Young, de Leeuw and Takane (1976). De Leeuw and van Rijckevorsel (1980) and Gifi (1982) use another loss function, for two reasons: (i) the treatment of missing data becomes more simple, (ii) both variables with a single quantification or transformation, and variables with multiple quantifications can be analysed simultaneously: a combination of NCA and MCA.

This alternative loss function is given by:

$$\sigma_{\text{alt}}(X, a, y) = m^{-1} \sum_j SSQ(G_j y_j a_j' - X). \tag{1.23}$$

In this function the rank-one matrix $y_j a_j'$ can be replaced for certain variables by a matrix Y_j, not necessarily of rank one, in order to treat variables with multiple quantifications.

Both loss functions (1.22) and (1.23) give the same results.

$$\sigma(X, a, y) = m^{-1} \sum_j a_j' a_j + 1 - 2m^{-1} \sum_j a_j' X' G_j y_j$$

$$\sigma_{\text{alt}}(X, a, y) = m^{-1} \sum_j \operatorname{tr}(a_j a_j') + p - 2m^{-1} \sum_j \operatorname{tr}(X' G_j y_j a_j')$$

$$= m^{-1} \sum_j a_j' a_j + p - 2m^{-1} \sum_j a_j' X' G_j y_j.$$

So that $\sigma_{\text{alt}} = \sigma + (p-1)$. Thus it is evident that both loss functions attain minima for the same X, A, Y (for missing data the solutions will usually not be the same). Again we use an interative procedure

(1) Take a matrix X of order $(n \times p)$, with $X'X = I$ and $u'X = 0$.
(2) Project the column vectors of X onto the subspaces G_j. The projections are $G_j Z_j = G_j D_j^{-1} G_j' X$, and

$$SSQ(G_j y_j a_j' - X) = SSQ(G_j Z_j - X) + SSQ(G_j y_j a_j' - G_j Z_j)$$

For multiple quantifications, the second term on the right-hand side can be made equal to zero by taking $Y_j = Z_j$. For nominal data the solution for y_j and a_j' can be found by taking the dominant singular value solution of $D_j^{1/2} Z_j$. This solution can also be found by an iterative procedure, which can also be used to treat ordinal data: the so-called inner iterations.

(2a) Take a vector y_j with $y_j' D_j y_j = 1$.
(2b) Now project $G_j Z_j$ on $G_j y_j$. This gives b_j. Thus

$$G_j y_j b_j' = G_j y_j y_j' G_j G_j Z_j, \text{ and}$$

$$SSQ(G_j y_j a_j' - G_j Z_j) = SSQ(G_j y_j b_j' - G_j Z_j) + SSQ(a_j - b_j);$$

set $a_j = b_j$
(2c) Project the rows of Z_j onto a_j: $a_j z_j' = (a_j' a_j)^{-1} a_j a_j' Z_j'$,

and

$$SSQ(G_j y_j a_j' - G_j Z_j) = SSQ(G_j z_j a_j' - G_j Z_j) + a_j' a_j SSQ(G_j(y_j - z_j))$$

The minimum of the second term on the right-hand side can be found, depending on the type of data, by linear regression, monotonic regression, or simply by setting $y_j = z_j$. After normalization of $G_j y_j$, we can return to (2b) or go on to step 3. Before doing so, we may observe that $G_j z_j$ is in deviations from the mean:

$$u' G_j z_j = u' D_j Z_j a_j (a_j' a_j)^{-1} = u' D_j D_j^{-1} G_j' X a_j (a_j' a_j)^{-1}$$
$$= u' X a_j (a_j' a_j)^{-1} = 0.$$

(3) For fixed $G_j y_j a_j'$ we now have to find a new space X. If we let $Q_j = G_j y_j a_j'$, then we have to minimize the loss function

$$\sigma(X) = m^{-1} \sum_j (Q_j - X) \tag{1.24}$$

where $X'X = I$. This is a Procrustus problem (Cliff, 1966). Minimizing (1.24) is the same as maximizing,

$$\sum_j \text{tr}(X' Q_j) = \text{tr}(X' \sum_j Q_j) = \text{tr}(X' Q).$$

where $Q = \sum_j Q_j$. If we define the SVD-solution $Q = K \Lambda L'$ then

$$\text{tr}(X'Q) = \text{tr}(X' K \Lambda L') = \sum_i (l_i' X' k_i) \lambda_i.$$

As $k_i' X l_i \leq 1$, the above expression reaches a maximum for $k_i X l_i = 1$. This means $X = KL'$. As the vectors of Q_j, and thus the vectors of Q, are in deviations from the mean, also X is in deviations from the mean. We can now return to the second iteration step.

1.4.3. *The geometry of NCA*

For every variable we have p component loadings and the transformed variables can be mapped into a subspace or 'X-plane' as vectors Xa_j. As was the case for PCA, the squared component loadings $(a_{js})^2 (s=1, \ldots, p)$ can be interpreted as explained variances per variable. Usually these quantities are referred to as measures of 'single fit'. Analogous to MCA, the barycentra $Z_j = D_j^{-1} G_j' X$ can be mapped into the 'X-plane' by projecting the vectors $G_j D_j^{-1}$ onto the 'X-plane': $XZ_j' = XX'G_j D_j^{-1}$. And, also analogous to MCA, usually categories with low frequencies will have extreme positions, when they are represented by Z_j. Where MCA uses the term 'discrimination measure' for indicating how close the object scores are located around their barycentra, NCA uses 'multiple fit'.

As opposed to MCA the multiple fit is not maximized for all variables per dimension. It just indicates the dispersion of the categories per dimension.

However, for nominal data (with single quantifications) we do have a maximum dispersion of the categories in the direction of Xa_j. So we first have to project the object scores onto Xa_j, instead of x_1 and x_2 as was the case for MCA. The corresponding measure is the single fit $(a_{js})^2$ summed over all dimensions. For all variables together:

$$\sum_i \psi_i^2 (\psi_i^2\text{: latent roots of the correlation matrix: } i=1, \ldots, p).$$

For ordinal data also the order of the categories has to be correct, so that the single fit might be smaller than a measure of dispersion of the barycentra (when projected on Xa_j) would indicate.

2. A REFORMULATION

In this section we reformulate multiple correspondence analysis and non-metric principal component analysis in terms of optimal scaling.

2.1. Optimal scaling

Linear multivariate analysis can be generalized in several directions to non-linear multivariate analysis. An important generalization is given by non-linear multivariate analysis using optimal scaling, where optimality is defined in terms of the correlation matrix of scaled (or quantified or transformed) variables. We could describe this generalization as follows. Suppose we are free to choose m elements (vectors or random variables) q_j from m subsets L_j of a linear space L. For every choice $(q_1, \ldots, q_m)$, $q_j \in L_j (j=1, \ldots, m)$, we can compute a correlation matrix $R(q_1, \ldots, q_m)$ (we do not consider those $(q_1, \ldots, q_m)$ for which this computation is impossible). We also have an objective function μ, defined on the set of all possible correlation matrices. Our non-linear multivariate analysis technique then consists of choosing the $q_j \in L_j$ in such a way that the function $\mu(R(q_1, \ldots, q_m))$ is maximized. More generally, we could say that we are interested in computing some, or all, of the stationary points of $\mu(R(q_1, \ldots, q_m))$ on $L_1 \times \ldots \times L_m$.

In defining our technique this way, we must consider two aspects: the form of the subspaces L_j and the nature of the objective function. Different choices in respect of one or both aspects will usually result in different techniques of analysis. The question is whether different choices would, in special situations, yield the same results. In the following sections we hope to give an answer for this question, in particular for two different techniques: MCA and NCA. Before presenting these two techniques in terms of optimal scaling, we first give some examples of possible subspaces L_j and objective functions μ.

Optimal scaling may be used to compute quantifications for observations that are missing. In this case the subspaces L_j are formed by elements with values that equal the observed values but which are arbitrary for observations that are missing. Using these subspaces, we can find optimal quantifications for the missing data (cf. Wold and Lyttkens, 1969).

Another application is given by the analysis of ordinal data. In this case the subspaces L_j are convex cones: if $q_j, p_j \in L_j$, and $\alpha > 0$, then $\alpha q_j \in L_j$ and $q_j + p_j \in L_j$. These subspaces are applied by Kruskal (1965), Kruskal and Shepard (1974), de Leeuw, Young and Takane (1976), Young, de Leeuw and Takane (1976), Young, Takane and de Leeuw (1978), and many others. Of course, for nominal data, we can also form subspaces. Again each variable corresponds to a subspace and an element corresponds to an arbitrary quantification of the categories of a variable. Obviously these spaces are linear. This definition too has been used in the articles of de Leeuw, Young and Takane mentioned above. Other choices of subspaces are polynomials of a certain degree, or polynomial splines (van Rijckevorsel, 1987). See also Chapters 2 and 3.

As regards the objective function we first mention the situation where there exists a partitioning of the variables into two sets. As objective function we may then use the canonical correlation; special, that is linear, applications are given by Anova, Manova, discriminant analysis and multiple regression (cf. van de Geer, 1986). When there is no prior partitioning of the variables into subsets, we treat the variables in a symmetric way and the objective function will then be invariant under permutations of the variables. Such a function is given for example by the sum of the correlations (Horst, 1965). Another class of objective functions is given by the symmetric functions of the latent roots (ψ_i) of the correlation matrix. We mention the determinant of the correlation matrix ($\prod \psi_i$), used by Chang and Bargmann (1974) and the sum of squares of the correlations ($\sum \psi_i^2$), suggested by Kettenring (1971). Another important function is the largest latent root of the correlation matrix, proposed by Horst (1965) and Carroll (1968).

The most popular programs for non-linear principal component analysis use the sum of the p largest latent roots of the correlation matrix as their objective function. Of course, maximizing the sum of the p largest latent roots is equivalent to minimizing the sum of the $m-p$ smallest latent roots. We shall denote these particular functions as μ_p, with the understanding that different choices of p result in different analyses.

2.2. Another definition of NCA

If we denote the intersection of a subspace L_j and the unit sphere S by L_jS, then all elements $q_j \in L_jS$ are normalized to unit length and for each choice $(q_1, \ldots, q_m)$, we can define a matrix $R(q_1, \ldots, q_m)$ with elements $r_{jl}(q_1, \ldots, q_m) = q_j'q_l$, the ordinary inner product of q_j and q_l. The matrix $R(q_1, \ldots, q_m)$ can be regarded as a correlation matrix, in the sense that it is positive semi-definite and that it has diagonal elements equal to one. The problem of non-metric principal component analysis (NCA) is to find $y_j \in L_jS$ in such a way that the sum of the p largest latent roots of the matrix $R(q_1, \ldots, q_m)$ is maximized. More generally, we shall be interested in all solutions of the stationary equations corresponding to the maximization of $\mu_p(R(q_1, \ldots, q_m))$. We consider the special case where the subspaces L_j are linear and of finite dimension k_j.

If the dimensionality of the total space L equals n, then orthonormal bases of $L_j (j=1, \ldots, m)$ can be depicted by matrices F_j of order $(n \times k_j)$, and $F_jw_j \in L_jS$ if and only if $w_j'w_j = 1$. If we set $B_{jl} = F_j'F_l$, then,

$$r_{jl}(q_1, \ldots, q_m) = w_j'B_{jl}w_l.$$

We collect the matrices F_j in a supermatrix $F = (F_1, \ldots, F_m)$. The matrix $B = F'F$ is called the Burt table, named after Burt (1950). This latter matrix is of course dependent on the choice of a basis. If we define W as the direct sum $W = w_1 \# \ldots \# w_m$, so that $W'W = I$, then clearly,

$$R = W'BW, \tag{2.1}$$

$$\mu_p(R(q_1, \ldots, q_m)) = \mathrm{tr}(A'RA), \tag{2.2}$$

where the matrix A varies over all matrices of order $(m \times p)$, for which $A'A = I$. For a maximum of (2.2) the vectors of A must be latent vectors of R. If we write $\sigma_p(L_1, \ldots, L_m)$ for a maximum of $\mu_p(R(q_1, \ldots, q_m))$ with $q_j \in L_jS$, then,

$$\sigma_p(L_1, \ldots, L_m) = \max\{\mathrm{tr}(A'W'BWA)\} = \max\{\mathrm{tr}(T'BT)\}, \tag{2.3}$$

with A and W varying over matrices of the prescribed form, or T varying over matrices $T = WA$ of order$(\sum k_j \times p)$, for which $T'T = I$ and T consists of matrices $T_j = w_ja_j'$ of order $(k_j \times p)$, where a_j is row j of matrix A. Thus T is blockwise of rank one; each subspace L_j defines a block. The maximum (2.3) is found by differentiation. We first write:

$$\operatorname{tr}(A'W'BWA) = \sum_l \sum_j \sum_s a_{js} w_j' B_{jl} w_l a_{ls}$$

$$= \sum_l \sum_j \gamma_{jl} w_j' B_{jl} w_l, \qquad (2.4)$$

where γ_{jl} are elements of $\Gamma = AA'$. A maximum subject to $w_j'w_j = 1$ is found by differentiating the function

$$\sum_l \sum_j \gamma_{jl} w_j' B_{jl} w_l - \sum_j \lambda_j (w_j' w_j - 1),$$

where λ_j are Lagrange multipliers. The maximum is found for

$$w_j' w_j = 1,$$

$$\sum_l \gamma_{jl} B_{jl} w_l = \lambda_j w_j.$$

As mentioned above, for a maximum of (2.4) the matrix A must comprise latent vectors of R, and thus the stationary equations are given by

$$RA - A\Omega, \qquad (2.5)$$

$$\sum_l \gamma_{jl} B_{jl} w_l = \lambda_j w_j, \qquad (2.6)$$

where Ω is a diagonal matrix of order $(p \times p)$ comprising latent roots of R.

From the equations, we can derive a relation between Ω and λ_j. An element of R is given by $r_{jl} = w_j' B_{jl} w_l$. An element of AA' is given by γ_{jl}. Both are symmetric matrices of order $(m \times m)$, so that for a diagonal element of RAA', we have on the one hand, according to (2.6),

$$\sum_l \gamma_{jl} w_j' B_{jl} w_l = \lambda_j,$$

on the other hand, according to (2.5),

$$RAA' = A\Omega A'.$$

Thus, we must have that λ_j is a diagonal element of $A\Omega A'$. Consequently,

$$\sum_j \lambda_j = \operatorname{tr}(\Omega), \quad \text{or} \quad \operatorname{tr}(\Lambda) = \operatorname{tr}(\Omega).$$

Although (2.5) and (2.6) can be used to construct convergent algorithms, they give little insight into the mathematical structure of the NCA problem.

It is not clear how many solutions there are of (2.5) and (2.6) nor is it clear how these various solutions might be related. However, there is one fortunate exception: $p = 1$.

2.3. Another definition of MCA

In the special case where $p=1$, the solution of (2.5) and (2.6) becomes much simpler and the problem of multiple correspondence analysis (MCA) is to compute some or all solutions. If $p=1$, the matrix T comprises one vector only and the restriction that the blocks of T must be of rank one is trivially satisfied. The solution is then found by maximizing $t'Bt$ subject to $t't=1$, and this is simply a latent root problem, leading to the equation,

$$Bt=t\mu. \tag{2.7}$$

Of course, we could decompose t into subvectors $t_j=w_ja'_j$, where a'_j is a scalar $a_j=(t'_jt_j)^{1/2}$, and $w_j=t_j(t'_jt_j)^{-1/2}$ (if $t'_jt_j=0$, then w_j is arbitrary with $w_j'w_j=1$). The relations between w and λ_j in (2.5) and (2.6) are then as follows:

$$\lambda_j=\sum_l a_ja_lw'_jB_{jl}w_l=\sum_l t_j'B_{jl}t_l=t'_jt_j\mu,$$

and

$$\sum_j \lambda_j=\Omega.$$

Hence $\Omega=\mu$. If μ is the largest, or smallest, latent root of B, then μ must also be the largest, or smallest, latent root of R. For intermediate latent roots of B, we can only state that they must also be latent roots of the corresponding matrix R.

In order to emphasize the relationship between NCA and MCA, we could define the latter as the maximization of,

$$\rho_p(L_1, \ldots, L_m)=\max.\{\mathrm{tr}(V'BV)\}, \tag{2.8}$$

where V varies over matrices of order $(\sum k_j \times p)$ and $V'V=I$. The main difference with NCA, (2.3), is that the blocks of V need not be of rank one. As a consequence and as contrasted with NCA, MCA is nested, i.e. the solution for $p=k$ is the same as the first k solutions found for $p>k$. In addition, a MCA solution, according to (2.8), generates p correlation matrices, whereas a NCA solution generates only one correlation matrix. Clearly, we have,

$$\rho_p(L_1, \ldots. L_m)\geq\sigma_p(L_1, \ldots, L_m), \tag{2.9}$$

with the equality holding if and only if the p dominant latent vectors of B have blocks of rank one.

3. CORRESPONDING MCA AND NCA SOLUTIONS

In this section, we shall derive a condition under which both MCA and NCA find the same solution. The emphasis will be on the possible interpretations of this situation.

3.1. A special condition

At the end of the previous section, we gave a condition under which the maxima (2.3) and (2.8) are equal: $\rho_p=\sigma_p$. The rigidity of this condition can be lessened by demanding that a solution of (2.3) equals a stationary value of (2.8). Then we can still say that the two solutions are equivalent. This new condition means that there are latent vectors of B which are block-wise of rank one, but which need not be necessarily the latent vectors corresponding to the p largest latent roots of B. In particular, the situation for which there are m such latent vectors is interesting, because then the relation between MCA and NCA holds for any p.

Therefore we assume that there are m latent vectors of B which can be decomposed in the manner of equation (2.3) to give: $V=WA$, $W=w_1 \# \ldots \# w_m$, $W'W=I$ and $A'A=I$ (because A is of order $(m\times m)$, we must also have $AA'=I$). This means that,

$$BWA=WA\Omega, \tag{3.1}$$

where Ω is again a diagonal matrix comprising latent roots. For these m different MCA solutions (by which we mean separate latent vector solutions and not m solutions of (2.8)), we find only one correlation matrix.

$$R=W'BW.$$

We already know that each MCA latent root corresponds to a latent root of the associated correlation matrix. In this case all latent roots of R are latent roots of B:

$$RA=W'BWA=W'WA\Omega=A\Omega. \tag{3.2}$$

We can now derive an important relation,

$$BW=BWAA'=WA\Omega A'=WRAA'=WR,$$

so that,

$$BW=WR, \quad \text{or,} \quad B_{jl}w_l=w_jr_{jl}. \tag{3.3}$$

Conversely, when there are vectors $w_j (j=1, \ldots, m)$ satisfying (3.3), we can combine these vectors with p latent vectors of R, in order to form p latent vectors of B which are block-wise of rank one. For,

$$BWA=WRA=WA\Omega.$$

Although we can form m such latent vectors of B, it follows from (3.1) that $\rho_p (p<m)$ has a stationary value. Also σ_p has a stationary value. This follows, in particular, from the fact that the stationary equations (2.5) and (2.6) are satisfied:

$$(2.5): \quad RA=A$$

$$(2.6): \quad \sum_l \gamma_{jl}B_{jl}w_l=\sum_l \gamma_{jl}w_jr_{jl}=w_j\Big(\sum_l \gamma_{jl}r_{jl}\Big)=w_j\lambda_j.$$

Thus, if condition (3.3) is satisfied we can form m different MCA solutions and $C(m, p)$ stationary NCA solutions (using definition (2.8). We could also say that there are $C(m, p)$ stationary MCA solutions).

Besides MCA and NCA, condition (3.3) can also be related to other techniques of analysis. In fact, all (differentiable) functions $\mu(R(q_1, \ldots, q_m))$ have a stationary value if condition (3.3) is satisfied. A stationary value of $\mu(R(q_1, \ldots, q_m))$, where $q_j = F_j w_j$ is subject to $w_j' w_j$, is found by differentiating.

$$\mu(R(q_1, \ldots, q_m)) - \sum_j \lambda_j (w_j' w_j - 1), \tag{3.4}$$

where λ_j are Lagrange multipliers. As $r_{jl} = w_j' B_{jl} w_l$, we can use the chain-rule to find:

$$w_j' w_j = 1,$$

$$\sum_l \frac{\partial \mu}{\partial r_{jl}} B_{jl} w_l = \lambda_j w_j. \tag{3.5}$$

Clearly, if condition (3.3) is satisfied, then (3.5) is also satisfied, i.e. all functions $\mu(R(q_1, \ldots, q_m))$ have a stationary value. Of course, this is due to the fact that $R(q_1, \ldots, q_m)$ itself is stationary if (3.3) is satisfied. Conversely, if $R(q_1, \ldots, q_m)$ is stationary then (3.3) is satisfied.

Condition (3.3) corresponds to another desirable property. Suppose v is a MCA solution, and thus a latent vector of B, but not one of the m solutions that can be formed by using (3.3). As the latent vectors of B are orthogonal (when the latent roots are equal, they can be chosen to be orthogonal), we may write,

$$v' WA = (0, \ldots, 0), \quad \text{or,} \quad v' WAA' = (0, \ldots, 0), \quad \text{or}$$

$$v' W = (0, \ldots, 0), \quad \text{or,} \quad v_j' w_j = 0. \tag{3.6}$$

This means that the quantifications are orthogonal for each variable separately. In the terminology suggested by Dauxois and Pousse (1976), the solutions are not only weakly orthogonal, because they are latent vectors of B, but actually strongly orthogonal. Thus $F_j v_j$ and $F_j w_j$ are orthogonal, but $F_j v_j$ is also orthogonal to the transformations of the other variables $F_l w_l (l = 1, \ldots, m)$. For,

$$v_j' F_j' F_l w_l = v_j' B_{jl} w_l = v_j' w_j r_{jl} = 0.$$

This means that the space Z spanned by the vectors $F_j w_j (j = 1, \ldots, m)$ is orthogonal to the transformed variables of all other MCA solutions. In particular this holds for all other solutions of (3.3):

$$B_{jl} v_l = v_j s_{jl}.$$

If all subspaces L_j have the same dimensionality k, then the maximum number of solutions of (3.3) equals k. We can depict the condition for the existence of these k solutions as follows. If we collect the solutions $w_{js} (s = 1, \ldots, k)$ in an orthonormal matrix K_j, and the correlations r_{jl} for the k different solutions in a diagonal matrix R_{jl}, then the total condition is given by:

$$B_{jl}K_l = K_j R_{jl}. \tag{3.7}$$

If the dimensionalities of the subspaces are different, we can maintain the formulation of (3.7), with the understanding that the matrices K_j are of order $(k_j \times k_j)$, where k_j is the dimensionality of subspace L_j, and R_{jl} is a matrix of order $(k_j \times k_l)$ having non-zero elements in the positions (i, i) where ($i = 1, \ldots,$ min (k_j, k_l)).

A solution of (3.7) for which all submatrices B_{jl} are diagonalized simultaneously, generates a number of orthogonal spaces $Z_s (s = 1, \ldots, \max(k_j))$. Each space Z_s is spanned by m_s vectors: those transformed variables for which $k_j \geq s$. Each space generates m_s separate MCA solutions and $C(m_s, p)$ NCA solutions (if $m_s < m$, we may use arbitrary quantifications for variables with $k_j < s$, while, at the same time, the weights a_j equal zero; for then the stationary equations are still satisfied). This amounts to $\sum m_s = \sum k_j$ different MCA solutions.

Before discussing the interesting interpretations of the conditions (3.3) and (3.7) we would like to make a remark on the possible nestedness of the NCA solutions. As we have already observed, the NCA solutions are not nested in general, i.e. correlation matrices formed at a maximum of σ_p are different for different p. If (3.7) is satisfied then the correlation matrix found at a maximum of σ_k also generates stationary values of σ_p, where $p \neq k$. However, it is not all that clear whether this correlation matrix corresponds to a maximum of σ_p, $(p \neq k)$.

With respect to the MCA solutions we can say that, although it is not impossible, the largest p latent roots are generally not generated by the same correlation matrix. Only if the p largest latent roots come from the same correlation matrix, do we have $\rho_p = \sigma_p$ (see (2.9)).

3.2. A general interpretation

If we represent (3.3) in words, it says that if two of the matrices C_{jl} have a subscript in common, then they have a singular vector in common. Condition (3.7) then says that they share all singular vectors corresponding to the common index. This is of course a very strong condition which greatly restricts the form of the matrices. The point of interest at the moment, however, is the interpretation. One possible interpretation says that the (normalized) quantifications of the variables remain invariant under removal of other variables from the analysis.

As far as we are considering the (normalized) quantifications, we could say that the analyses are nested with respect to the variables. This also holds for the condition (3.3) if we restrict our attention to the MCA and NCA solutions generated by this single system.

Obviously this situation always exists for numerical data. In that case the linear subspaces are one-dimensional and the condition (3.3) is a trivial one. We also expect approximations of this ideal situation to be much better for ordinal data than for nominal data, the former being more 'one-dimensional' than the latter. We could say that the nominal variables measure more than one thing. Consequently, in a homogeneity analysis, we expect a nominal variable to act upon the other variables in two ways, the first of which is 'how well is something measured', the second is 'what is measured'. Therefore, we do not expect the conditions (3.3) or (3.7) to be satisfied, or even nearly satisfied for nominal data: variables will measure different things before and after removal of a variable from the analysis, that is to say, the (normalized) quantifications will be different.

Another interpretation is given by a reformulation of (3.3) in terms of projections. We can rewrite (3.3) as

$$F_jF_j'F_lw_l = F_jw_jr_{jl}. \tag{3.8}$$

This indicates that the projection of the transformed variable, $q_l = F_lw_l$, onto the subspace L_j, spanned by the orthonormal basis F_j, coincides with the projection on the one-dimensional space spanned by $q_j = F_jw_j$. This holds for all mutual projections. As a consequence, the projections of all vectors in the space Z (spanned by $q_l = F_lw_l(l = 1, \ldots, m)$) onto the subspaces L_j will coincide with the projections on the one-dimensional spaces q_j. Conversely, projections of vectors in L_j onto Z coincide with projections on q_j. Namely, if we denote a vector in L_j by Fa, where a is a vector containing zeros with the exception of those k_j elements that correspond to the k_j vectors of F_j, then the projection of Fa onto the space Z, spanned by FW, is given by:

$$\begin{aligned} FW(W'BW)^{-1}W'F'Fa &= FWR^{-1}W'Ba = FWR^{-1}RW'a \\ &= FWW'a = F_jw_j(w_j'a_j), \end{aligned} \tag{3.9}$$

which is a vector in the space spanned by $q_j = F_jw_j$. In the next subsection, we shall see that these projections have a special meaning if we use bases of indicator functions. Then we also have special interpretations of the conditions (3.3) and (3.7).

3.3. The interpretation for bases of indicator functions

In the first section, where we used indicator functions, we derived a relation between the optimal MCA-quantifications y_{js} and the object scores:

$$y_{js} = D_j^{-1}G_j'x_s.$$

This indicated that the projection of x_s on the subspace spanned by G_j coincides with the projection on the one-dimensional space of the transformed variable G_jy_{js}. As the vectors of G_j are indicator functions, we have a special interpretation for the operator $D_j^{-1}G_j'$. Variable h_j partitions the objects, or individuals, into k_j subgroups, indicated by the categories of variable h_j. The number of objects in these groups can be found in the diagonal of $D_j = G_j'G_j$. The averages of a variable, for instance x_s, for these k_j groups can then easily be formed by:

$$D_j^{-1}G_j'x_s.$$

The averages y_{js} lie on the regression line of the linear regression of x_s on y_{js} (or G_jy_{js}). In fact, linear regression means projecting a vector on another vector, and a non-linear regression could be defined as the projection of a vector on the space of all non-linear transformations of another vector. In the present case of discrete data, non-linear regression means projecting onto a subspace L_j or G_j. If these two regressions, linear and non-linear, coincide, we say that the regression is linearized. In the present case this means that no other transformation of h_j, than G_jy_{js}, gives better predictions of x_s. The variance between the groups can be fully explained by a linear function of G_jy_{js}.

For stochastic variables the 'percentage of variance between groups' is expressed in the correlation ratio:

$$\text{c.r.}(\mathbf{x}_1\mathbf{x}_2) = \frac{\text{Var}(E(\mathbf{x}_1|\mathbf{x}_2))}{\text{Var}(\mathbf{x}_1)}, \tag{3.10}$$

which is the ratio of the variance of the conditional expectation of $\mathbf{x}_1$ given $\mathbf{x}_2$, and the variance of $\mathbf{x}_1$. If the conditional expectation lies on the regression line, then the correlation ratio equals the squared correlation r^2. For discrete variables the correlation ratio can be simply related to the projection on a subspace G_j.

Returning to MCA, the relations between the transformed variables can be formed by using (1.10):

$$\sum_l D_j^{-1}C_{jl}y_l = y_j\psi^2. \tag{3.11}$$

This indicates that the average regression of y_l on y_j is linear: the summation of the projections of all $G_ly_l(l=1, \ldots, m)$ on G_j is in the one-dimensional space of G_jy_j. Let us now consider the situation where (3.3) or (3.7) is satisfied. The Burt table is given by:

$$B = D^{-1/2}CD^{-1/2}. \tag{3.12}$$

If we normalize the transformed variables, $y_jD_jy_j = 1$, then y_j can be written as $y_j = D_j^{-1/2}w_j$. Condition (3.3) is then given by:

$$D_j^{-1}C_{jl}y_l = y_jr_{jl}. \tag{3.13}$$

We now see that all regressions between the transformed variables are linearized. Hence, all correlation ratios equal the squared correlations. These findings are important because now the correlation matrix accounts for the whole bivariate relationship amongst the transformed variables, whereas it usually only gives linear relations.

4. SOME THEORETICAL EXAMPLES

Some reflection shows that the ideal conditions (3.3) and (3.7), which we discussed in the previous section, will usually not be met in practice. Very often the ideal situation can only be approximated. However, in this section we shall discuss some examples for some of which the conditions underlying this ideal situation are always met.

4.1. The trivial solutions

When discussing MCA in the first section, we observed that indicator functions, used as bases of the subspaces of non-linear transformations, generated meaningful solutions, as well as a number of trivial solutions. As we shall see, these trivial solutions have a quite natural place within the framework of the previous section. In fact, they give a trivial example where the condition (3.3) is always met.

We have already noticed that, by using indicator functions, condition (3.3) can be transformed as follows.

$$D_j^{-1}C_{jl}y_l = y_j r_{jl}.$$

This condition has a trivial solution given by: $r_{jl}=1$ and $y_j = un^{-1/2}(j, l = 1, \ldots, m)$ where the elements of u are units. All trivially quantified variables $G_j un^{-1/2} = un^{-1/2}$ have a length equal to unity and they coincide completely. They span a one-dimensional space Z_0. All elements of the so-called correlation matrix are units:

$$R = Y'CY = uu' \qquad (Y = y_1 \# \ldots \# y_m) \tag{4.1}$$

This matrix has one latent root equal to m, and $(m-1)$ latent roots equal to zero. These are the trivial MCA latent roots. The consequent strong orthogonality implies that the space Z_0 is orthogonal to all other transformed variables. Thus all other, meaningful transformed variables are in deviations from the mean, and the matrices R_s can be regarded as correlation matrices in the usual sense.

4.2. *Analyse des correspondances:* $m=2$

If we apply MCA to a data set of only two variables, the supermatrix of bimarginals has only one contingency table as submatrix. It is well known that, if we apply a similar analysis to this contingency table instead of to the supermatrix of bimarginals, the solutions agree up to a normalization and the latent roots can be directly related to one another (cf. Gifi, 1981a). This technique of analysis is called *analyse des correspondances* and it is discussed by many authors; see for instance Benzécri *et al.* (1973), Nishisato (1980), Gifi (1981a) or Greenacre (1984).

In this case of two variables, condition (3.7) is always met. In fact, when all submatrices B_{jl} can be diagonalized simultaneously, then condition (3.7) is satisfied:

$$K_j'B_{jl}K_l = R_{jl},$$

and this will always be possible when there are two variables, since then only one submatrix B_{12} has to be diagonalized. In that case a solution for (3.7) can be found easily by taking singular vectors of B_{12}. The orthogonal spaces Z_s are now spanned by two vectors, and the corresponding correlation matrices R_s have only one subdiagonal element which equals the singular value r_{12} of B_{12}. The MCA latent roots are consequently one plus the singular value and one minus the singular value.

In the case of *analyse des correspondances*, the regressions between the transformed variables are always linearized and the quantifications are always strongly orthogonal. Of course, in this case, NCA is not very meaningful.

4.3. Dichotomous variables: $k_j = 2$

For dichotomous or binary variables, the regressions are by definition linear, because a straight line can always be drawn through two points. If $k_j = 2$, the subspaces L_j are two-dimensional. However, one dimension is due to the trivial solution Z_0, which means that the variables can be quantified in deviations from the mean in only one direction. The frequencies directly induce a quantification, and similar to the case of numerical data, there is nothing to quantify. We simply compute the product-moment correlations (phi-coefficients, or point-correlations) and perform a PCA of the correlation matrix.

4.4. Normally distributed variables

Hitherto, we have assumed the dimension of a subspace L_j to be finite. In this section we shall consider subspaces of infinite dimensions. We have seen that the indicator functions are a perfectly satisfactory basis if L_j is the space of all non-linear transformations of a discrete variable which assumes only a finite number of values. However, if the number of categories of the variables is very large and close to the number of observations, difficulties might arise. In that case the use of indicator matrices is no longer satisfactory: for they will be close to a permutation of the identity matrix. Next to G_j, this will also be true for the matrices D_j and C_{jl}. Consequently, all latent roots of the multiple correspondence problem will be close to either zero or one, and the latent vectors will be very unstable and rather uninteresting.

This solution occurs for continuous variables. For, in practice, dealing with continuous variables really means dealing with discrete variables with a very large number of categories, close to the number of observations. In these cases the space of all non-linear transformations is simply too big, because it is approximately equal to the whole space L, and if each L_j is approximately equal to L then non-linear PCA does not make sense. Thus, we want to approximate the space of all non-linear transformations by using small-dimensional subspaces. Indicator functions correspond with approximations of non-linear functions by step functions. In the theory of the approximation of functions, it is well known that step functions give poor approximations of continuous or smooth mappings. A classical alternative is given by polynomials. Although polynomials have many attractive theoretical properties, they are not very suitable for approximating general continuous functions. The basic problem is that polynomials are too rigid; if we change a coefficient then the whole polynomial changes. Therefore polynomials of a very high degree are needed for the approximation of functions which do not behave in the same way over the whole range. A more satisfactory alternative is given by the so-called B-splines (De Boor, 1978). We shall not go further into this matter, this being dealt with in the Chapters 2 and 3. However, we will give an example in which polynomials are used. The theoretical example of this section is given by the multinormally distributed variables, for which it is known that all bivariate regressions are linearized, that is to say, condition (3.7) is satisfied. Suppose the random variables $\mathbf{x}_j$, $(j=1, \ldots, m)$ are jointly multivariate normal, with zero means, unit variances and correlations ρ_{jl}. The subspaces of non-linear transformations are given by:

$$L_j=\{\phi_j(\mathbf{x}_j)\,|\,\mathrm{var}\{\phi_j(\mathbf{x}_j)\}<\infty\}. \tag{4.2}$$

The mapping ϕ_j are assumed to be measurable. As a basis for L_j we use Hermite–Chebyshev polynomials ψ_ν (.), of degree $\nu=0, 1, \ldots$ which are orthonormal on the normal distribution:

$$\int \psi_\nu(x)\psi_\chi(x)N(x)\,\mathrm{d}x=\delta(\nu, \chi) \tag{4.3}$$

where $\delta(\nu, \chi)$ is the Kronecker delta. We can now expand $\phi_j(\mathbf{x}_j)$ as,

$$\phi_j(\mathbf{x}_j)=\sum_\nu a_{j\nu}\psi_\nu(\mathbf{x}_j). \tag{4.4}$$

For the covariance of $\phi_j(\mathbf{x}_j)$ and $\phi_l(\mathbf{x}_l)$ we have,

$$\mathrm{cov.}\,(\phi_j(\mathbf{x}_j), \phi_l(\mathbf{x}_l))=\sum_\nu\sum_\chi a_{j\nu}a_{l\chi}\mathrm{cov.}(\psi_\nu(\mathbf{x}_j), \psi_\chi(\mathbf{x}_l)). \tag{4.5}$$

The following identity is due to Mehler (cf. Lancaster, 1969),

$$\mathrm{cov.}\,(\psi_\nu(\mathbf{x}_j), \psi_\chi(\mathbf{x}_l))=\delta(\nu, \chi)\,(\rho_{jl})^\nu, \tag{4.6}$$

where ρ_{jl} is the correlation between $\mathbf{x}_j$ and $\mathbf{x}_l$. Clearly then,

$$\mathrm{cov.}\,(\phi_j(\mathbf{x}_j), \phi_l(\mathbf{x}_l))=\sum_\nu a_{j\nu}a_{l\nu}(\rho_{jl})^\nu. \tag{4.7}$$

With respect to (4.6) we could say, in the terminology of the previous section, that polynomials of the same degree, $\psi_\nu(\mathbf{x}_l), \ldots, \psi_\nu(\mathbf{x}_m)$, span a space Z_ν, orthogonal to all other spaces Z_χ, spanned by polynomials of another degree, $\chi \neq \nu$. This means that we could as well have taken the polynomials of a certain degree directly as the (normalized) non-linear transformations. For then condition (3.7) is satisfied. The MCA and NCA solutions are again generated by the latent root solutions of the correlation matrices:

$$R^{(\nu)} a_\nu = \omega a_\nu, \qquad \nu = 1, 2, \ldots, \tag{4.8}$$

where $R^{(\nu)}$ has elements $(\rho_{jl})^\nu$. Thus, for every matrix $R^{(\nu)}$ we find m MCA solutions and $C\,(m,p)$ NCA solutions. It is implied by general results on Hadamard products (Styan, 1973) that the largest latent root of $R^{(1)}$ cannot be smaller than the largest latent root of $R^{(2)}$, which in its turn cannot be smaller than the largest latent root of $R^{(3)}$, etc. For the smallest roots the converse holds. The largest and smallest MCA latent roots are, consequently, the largest and smallest roots of $R^{(1)}$, and both correspond to linear transformations.

The second largest latent root is more of a problem. It can either be the second largest latent root of $R^{(1)}$, or the largest latent root of $R^{(2)}$. In the former case, both the first and second MCA dimensions correspond to linear transformations; in the latter case the first MCA dimensions corresponds to a linear transformation, whereas the second dimension corresponds to a quadratic transformation.

For homogeneous variables, the largest latent root of $R^{(1)}$ will be considerably larger than the second latent root of $R^{(1)}$, and usually also the largest latent root of $R^{(2)}$ will be larger than the second root of $R^{(1)}$. Thus, homogeneous variables approximating the normal distribution usually have mappings (into the first two dimensions) which approximate parabolas: horseshoes.

For general ordinal conditions for data matrices to have horseshoe mappings we refer to Schriever (1986). The question, however, is whether this second, quadratic transformation contributes to our knowledge about the relations amongst the variables. In his doctoral thesis de Leeuw (1973) writes: 'As pointed out by Bartlett [1953] and McDonald [1968], in the classical case we suppose, more or less implicitly, that the component scores are stochastically independent'. In the present case there is a simple non-linear relation between the components and, obviously, the dimensions are not independent. As we noticed before, correlations can only fully account for the bivariate relations amongst the variables if the regressions are linearized. In our case, the regressions between the linear transformations and the quadratic transformations are not linearized at all; they are 'parabolized'. The non-correlatedness of the dimensions says nothing about the non-relatedness of the dimensions. Thus, in case one finds a horseshoe mapping, it is best to consider only the correlation matrix of the first MCA solution, or NCA solution. See also van Rijckevorsel (1987).

The NCA solutions are always generated by one of the $R^{(1)}$, $R^{(2)}, \ldots$, matrices. In practice, it is usually the first matrix $R^{(1)}$, which is the original, non-transformed, correlation matrix.

We discussed some situations where (3.3) and (3.7) hold. Data analytically the situations in which these properties are approximately true are just as interesting. Often we do not know nor can we find out beforehand how the data are distributed or whether there exist any other dependencies. If simultaneous diagonalization is approximately true the correlations in the off-diagonal blocks must be small; the percentage of corresponding dimensions between the eigenvectors of $R^{(1)}, R^{(2)}, \ldots, R^{(q)}$ and the observed complete set of homogeneity analysis scores is a measure for the existence of the blockwise rank-one structure. One can compute the correlations between the approximately diagonalizing eigenvectors and the actual homogeneity analysis transformations. De Leeuw (1982) and Bekker (1983) respectively constructed an algorithm and a computer program (PREHOM) to this purpose. The practical value of the program is not that obvious, but an application provides a tangible illustration of the blockwise rank one approximation. To this purpose we show the following example by van Rijckevorsel (1987) using the Holmquist data.

Pathologists study biopsy slides of the uterine cervix in order to classify carcinoma *in situ* and related lesions into five classes:

1 = negative
2 = atypical squamous hyperplasia
3 = carcinoma *in situ*
4 = squamous carcinoma with early stromal invasion
5 = invasive carcinoma

cf. Holmquist, McMahan and Williams (1967).

The major decision to give treatment or not is usually based on the fact whether a slide belongs to the classes 1 or 2 (= no treatment) or to the classes 3, 4 or 5 (= treatment). If several pathologists study the same set of slides we would like their judgements to be consistent at least with respect to give treatment or not. In practice this is an ideal and unrealistic situation because there exists no absolute true scale of slides that is perfectly partitioned diagnostically. Even for the best pathologist in the world there occasionally exist serious doubts about the right classification of certain slides.

Slides and subsets are quantified by the parameters x and y respectively, and the problem is to find a common scale x for slides and scores y for categories such that the common scale is maximally consistent with all weighted judgements of each pathologist $G_j y_j$ simultaneously.

These data are of the rating scale type and they show a non-perfect horseshoe in the first two dimensions of a homogeneity analysis. We do not report on this analysis here. The eigenvalues with their approximations and the correlations between the corresponding eigenvectors per axis are shown in Table 1.1

Table 1.1. The eigenvalues of and the correlations between the homogeneity analysis solution and its block diagonal approximation of the Holmquist data

			Eigenvalues		
	θ_1	θ_2	θ_3	θ_4	θ_5
Actual	5.56	5.20	2.75	2.09	1.68
approximation	5.50	5.24	2.50	2.21	1.53
Correlations between the corresponding eigenvectors	0.92	0.92	0.84	0.82	0.87

Without any extra assumptions the axes of the homogeneity analysis have approximately a blockwise rank-one structure, at least in the first five dimensions. This is more or less to be expected because of the type of data, the number of variables, the number of categories and the occurrence of a horseshoe in the first two dimensions. We observed a particular bad fit of pathologist no. 6 in the preceding homogeneity analysis; this is also reflected in a bad approximation regarding the size of the off-diagonal correlations between dimensions for this pathologist (not displayed). However, it is more efficient to look at the percentage of overlap between corresponding observed and approximated dimensions per pathologist. This can be expressed in chi square per pathologist between observed and approximate transformations (Bekker, 1983) see Table 1.2. Pathologist no. 6 has the smallest overlap and is hence the least diagonizable. Tentatively this could mean that no. 6's response is not so regularly distributed, is less of a rating scale type or is less order dependent than the responses of the other pathologists. To locate this odd man out by block diagonalization or by testing for order dependence in another way than by ordinary homogeneity analysis is respectively too cumbersome or downright impossible (there exist no statistical tests for order dependence (cf. Schriever, 1986)).

Table 1.2. The percentage of chi square in corresponding dimensions of the Holmquist data

Pathologist							*Averaged*	
							1	94
2	97						2	97
3	93	98					3	96
4	98	98	97				4	96
5	98	97	96	96			5	97
6	81	93	97	90	95		6	92
7	100	98	93	98	99	96	7	98

Component and Correspondence Analysis
Edited by J.L.A. van Rijckevorsel and J. de Leeuw

Chapter 2

Fuzzy Coding and B-Splines

Jan L.A. van Rijckevorsel
Department of Statistics, TNO NIPG, Leiden, The Netherlands

1. INTRODUCTION

The contents of this chapter are mainly based on van Rijckevorsel (1987)

1.1. The indicator function

This chapter deals with the limitations of crisp coding and hence with the limits of representing data by an indicator matrix as defined in Chapter 1.

Let us evaluate the indicator matrix.

(1) crisp coding is *robust* in the sense that differences between values within a subset are ignored. This is a rather crude treatment, generally most suited for noisy data. It is also an extremely separated and contiguous coding that often generates well separated and contiguous representations which, because of their parsimony, are easy to interpret. This is an important reason for its positive appeal.
(2) The number of parameters per variable is much smaller than the number of different data values. This is, however, only truly an asset if a non-linear analysis is the aim.
(3) There exists a natural definition of a column point as a combination of row points, which is attractive both in terms of geometry and numerical ease.
(4) The bivariate indicator matrix has a straightforward relationship with the representation by a two-way contingency table.
(5) Crisp coding is extremely simple and generates sparse matrices that give great computational ease.
(6) Any form or shape of a transformation function can be approximated through the steps of crisp coding.

(7) Crisp coding permits non-linear relationships between variables.
(8) When using crisp coding all analysis results are invariant under one-to-one preliminary transformations of the data.

Weak points of the indicator representation are the following.

(1) The arbitrariness of location and spacing of intervals, e.g. categories, if we handle data with too many values. This can be particularly damaging when dealing with small samples. Reasons for a particular discretization, based on matter of content, are generally not, or not sufficiently, available. The distribution of marginals is greatly affected by the choice of intervals and is of direct consequence to any subsequent analysis.
(2) It is abundantly clear beforehand that, for some data, subsequent intervals are functionally related to each other. And although the stepfunctions are dense in L^2, provided the number of discretization points goes to infinity, the latter is no realistic option and the crisp coding does not accommodate for smoothness or order dependence.
(3) Often there exists a considerable uncertainty about the allocation of data values to either interval, if the data values are lying close to the value of the interval boundary point. This demands for a local smoothness around such points not available in the rigid indicator framework.

1.2. Notation and terminology

In this chapter we use the same notation as in Chapter 1 and because all the results apply to one variable at a time and because we do not discuss relationships between variables, we omit the variable subscript, i.e. $G=G_j$, $k=k_j$ etc.

The domain of a real valued variable h, bounded by two values a and b, $a \neq b$, is partitioned into a number of intervals between a and b. Each interval is bounded by two boundary points, called knots. Assume that the knot sequence $\{t\}$ consists of an increasing sequence of knots on $[a, b]$:

$$a \leq t_1 < t_2 < \cdots < t_{u-1} < t_u \leq b$$

A variable is real valued on $[a, b]$ exclusively and the number of data elements equals n. We distinguish between knots that are enclosed by other knots on both sides (left and right) and knots that meet another knot on one side (left or right) only; the former are called *interior knots* and their number equals r and the latter are called *exterior knots* and their number is equal to $u-r$. We prefer this notation because in some cases the exterior breakpoints have multiple exterior knots (=coinciding knots). Their multiplicity is equal to the degree of the coding function. The intervals between knots are mutually exclusive and exhaustive and their number is equal to $k=r+1$, while each interval includes the left-hand knot.

A piecewise coding function is a positive function between 0 and 1 on some contiguous part on $[a, b]$; outside this contiguous part the piecewise coding function is generally equal to zero, apart from some special cases that will be discussed separately. The positive part of the coding function consists of one or more different contiguous functional pieces, that may or may not be of the same degree. The functional pieces are not necessarily exactly defined on the intervals defined by $\{t\}$, i.e. a functional piece can be positive on only a part of an interval defined by $\{t\}$. If more than one functional piece is positive on the same interval or parts thereof, we say they overlap. Overlapping functional pieces always add up to 1.

Every piecewise coding function is represented by a column-vector $G_q(h)$ with elements $\{G_q(h_i)\}$ $(i=1, \ldots, n;\ q=1, \ldots, w)$. For every variable there exist w different piecewise coding functions represented by correspondingly w column-vectors $\{G_q(h)\}$, that are collected in a pseudo-indicator matrix (PIM): $G(n*w)$. Because all overlapping functional pieces add up to one, each row-sum of G is equal to one, i.e.

$$\sum_q G_q(h)=1$$

The proper way to denote the qth coding vector for the transformation of a variable is $G_q(h)$, simplified to G_q in this chapter. The dimension of a coded variable is equal to the number of coding vectors, i.e. the number of columns of G, needed for a complete coding of that variable.

A B-spline is a piecewise coding function with functional pieces of degree v that is positive on exactly $v+1$ consecutive intervals with an overlap of exactly v intervals with the next B-spline, all intervals defined by $\{t\}$. The number of interior knots $(=r)$ must be greater than or equal to v, while the number of B-splines, needed to code a variable with knot sequence $\{t\}$ is equal to $r+v$, i.e. $w=k+v$. The order of a B-spline is equal to $v+1$. The optimality properties of spline functions are discussed in Section 4.2.

The transformation function $\Phi(h)$ is now defined as:

$$\Phi(h)=\sum_q \alpha_q G_q(h),$$

i.e. Φ is a linear combination of piecewise coding functions. The dimensionality of Φ is equal to w and if the basis consists of coding functions of a degree maximally equal to v, the maximum degree of Φ is also equal to v. The smoothness of Φ can be evaluated by the existence of the $(v-1)$th derivative(s) at the interior knots. If G_q consists of a set of B-splines, Φ is defined as a spline function.

$\alpha_q (q=1, \ldots, w)$ are called the weights or coefficients of Φ. Provided the knot sequence and type of coding function are fixed, these coefficients are the only determining parameters of Φ. This means a considerable reduction of the number of parameters from originally n to w. The least squares estimation of $\{\alpha_q\}$ requires a (nearly) orthogonal set of basis functions, or, in other words a well conditioned basis.

A basis is orthogonal if the scalar-product between basis functions $G_q^T G_r = 0$ whenever $q \neq r$. A basis is nearly orthogonal if the matrix of such scalar-products is diagonally dominant, i.e. if the row (column) sum of any row (column) of the off-diagonal elements of such a matrix is smaller than, or equal to, the absolute value of the corresponding diagonal element, cf. Conte and De Boor (1980, p. 250):

$$|G_q^T G_q| \geq \sum_{l \neq q} |G_l^T G_q|, \qquad (q = 1, \ldots, w).$$

The near orthogonality of fuzzy coding is demonstrated in Section 3.

1.3. The state of the art

The drawbacks of the indicator matrix motivated several, predominantly French, authors to the generalization into other forms of coding that develop quite naturally. According to Benzécri, quoted by Le Foll (1979), the purpose of a coding is to represent in a reliable way relations, that occur in reality, by relations between mathematically defined variables; and in such a manner that this mathematical structure becomes simpler through computation; hence we obtain a simpler structure that has a greater intuitive and cognitive appeal while being mathematically consistent. This leaves enough freedom indeed to shop around for alternative coding systems.

Apparently it is Bordet (1973), who is the first one to introduce the notion of a fuzzy coding (= *codage flou*) on the empirical grounds of accommodating for the lack of local smoothness around the interval boundary points (= knots) of crisp coding. The term *codage flou* seems to be generally accepted in France for all kinds of alternative coding systems, to be used in various types of data-analysis techniques cf. Lafaye de Micheaux (1978), Le Foll (1979), Gallego (1980), Martin (1980, 1981), Gautier and Saporta (1982) and Mallet (1982). A practical definition of fuzzy coding is that we can code a variable with any arbitrary set of weights as long as they consist of real numbers and add up to one. Crisp coding is hence a special case of fuzzy coding. Note that by changing from crisp coding to fuzzy coding we implicitly use information on order and/or spacing of, at least some, data values.

Piecewise coding functions originate with Bordet (1973). Guitonneau and Roux (1977) and Ghermani, Roux and Roux (1977) propose and apply fuzzy codes on empirical grounds. See also Greenacre (1984, p. 159, 1981, p. 143).

Lafaye de Micheaux (1978) is the first to connect fuzzy coding, i.e. piecewise linear coding, to the theoretical work on the approximation of continuous non-linear analysis by increasingly finer steps by Dauxois and Pousse (1976).

Lafaye de Micheaux establishes several theoretical and empirical convergence properties of fuzzy coding as a functional approximation technique for canonical analysis with examples on simulated and real (medical) data.

Le Foll (1979) extends fuzzy coding to the coding of non-hierarchical graphs, proximity measures and polynomial coding and relates explicitly to multiple correspondence analysis.

Gallego (1980) exclusively discusses fuzzy coding in relation to multiple correspondence analysis in a rather pragmatic way, and uses the *metric* defined by piecewise linear coding as input for ordinary PCA and cluster analysis.

In Chapter 5 Martin focuses on the theoretical probabilistic properties of fuzzy coding and his work relates to chi square, error theory, non-observed random variables, probabilistic non-linear PCA, etc. In Chapter 3 de Leeuw and van Rijckevorsel discuss one kind of fuzzy coding: B-splines and the relationship to homogeneity analysis and non-metric PCA in particular.

Resuming, there exist three motives to study fuzzy piecewise coding

(a) *Convergence in sample size*. Crisp coding is extremely sensitive to sample fluctuations for small sample sizes (i.e. when n is small). Therefore we are interested in an alternative piecewise coding that is less sensitive while it converges to crisp coding for $n \to \infty$ (Martin, 1980, 1981; Ghermani, Roux and Roux, 1977; Guitonneau and Roux, 1977; Gallego, 1980).

(b) *Convergence in discretization*. The analysis of discretized variables is often interpreted as the approximation of an identical analysis of not coded variables. The alternative piecewise coding gives a better approximation in this way than crisp coding does. See Dauxois and Pousse (1976), Lafaye de Micheaux (1978), Martin (1981) and Mallet (1982).

(c) Crisp coding seems a rigid and arbitrary way to accommodate for a mathematical representation of a variable, cf. Benzécri's purpose of coding as quoted by Le Foll (1979).

2. LOCAL CODING

Piecewise coding functions can be ordered by their functional degree and amount of overlap. When ordered by their regularity, the crisp coding, or zero degree B-spline, comes first together with the semi-discrete coding function, which is also a uniform function with some extra steps around each interior knot. Next is the trapezoidal function that is linear in a restricted area around the knots, but equal to a uniform function elsewhere. The latter two types of fuzzy coding are discussed by Martin in Chapter 5. Two subsequent semi-discrete or trapezoidal coding functions overlap each other in the restricted area around each interior knot. More regular is the piecewise linear coding function, or first degree B-spline, that consists of two linear pieces, connected at a knot, and each of them is positive on exactly one interval. Two subsequent first degree B-splines have an overlap of exactly one interval. The semi-exponential coding is a piecewise linear coding function with exponential pieces defined on the extreme intervals. The quadratic coding function, or second degree B-spline, is a quadratic function defined on exactly three intervals, with an overlap of exactly two intervals with the next subsequent coding function. The existence of coding functions of a degree greater than two is well known. Their practical relevance for data analysis is however not convincingly clear. See Chapter 3.

In order to illustrate what the different coding functions look like and how they are stored in a pseudo-indicator matrix (PIM), we code a data-vector with 33 elements, representing the national public spending in the Netherlands in the period from 1951 until 1981, with a fixed knot sequence $\{t\}=\{30, 43, 54, 67\}$, cf. van Rijckevorsel and van Kooten (1985). See Table 2.1 and 2.2.

2.1. Crisp coding

This coding has many names: crisp coding, step function, nominal categorization, *codage par le mésure de Dirac*, indicator function, piecewise constant function, coding by dummy variables and the representation by a first order (=zero degree) B-spline. This binary representation of data is a hallmark of categorical data.
Definition:

$$g_{qi}=\begin{cases}1, & t_q \leq h_i < t_{q+1}\\ 0, & \text{elsewhere.}\end{cases}$$

The basis-matrix G consists of $w=r-1$ coding vectors G_q $(q=1, \ldots, w)$, with all rows adding up to one. There is no need to dwell upon this kind of coding here, it having been discussed in Chapter 1 already.

2.2. Coding by the first order semi-discrete function

This coding, first suggested by Ghermani, Roux and Roux (1977) as a form of fuzzy coding is defined by:

$$g_{qi}=\begin{cases}c_q, & t_q - v_q^{\mathrm{a}} \leq h_i < t_q + v_q^{\mathrm{b}}\\ 1, & t_q + v_q^{\mathrm{b}} \leq h_i < t_{q+1} - v_{q+1}^{\mathrm{a}}\\ 1-c_{q+1}, & t_{q+1} - v_{q+1}^{\mathrm{a}} \leq h_i < t_{q+1} + v_{q+1}^{\mathrm{b}}\\ 0, & \text{elsewhere,}\end{cases}$$

while v_q is called the fuzzy weight and v_q^{a} and v_q^{b} are the overlap constants around the knot t_q. The resulting function is identical with the indicator function over the larger part of an interval. However, if the data-value h_i is close to the knot between two intervals, another positive constant c_q which is smaller than one, is assigned to h_i and its complement is assigned to h_i as the coding for the next interval. The semi-discrete function converges to an ordinary stepfunction if

$(v_q^a + v_q^b) \to 0$ and/or $c_q \to 1$. The weak point of this form of coding is the arbitrariness of the fuzzy weights and the overlap constants. The occasion where we will have sufficient information to act upon will be rare indeed. See also Chapter 5.

2.3. Coding by first order trapezoidal functions

This coding originates with Guitonneau and Roux (1977) and is further discussed by Le Foll (1979), Martin (1980) and Greenacre (1981, 1984).

The definition of the trapezoidal function:

$$g_q i = \begin{cases} \dfrac{h_i - t_q - v_q^a}{v_q^a + v_q^b}, & t_q - v_q^a \leq h_i < t_q + v_q^b \\ 1, & t_q + v_q^b \leq h_i < t_{q+1} - v_{q+1}^a \\ \dfrac{t_{q+1} + v_{q+1}^b - h_i}{v_{q+1}^a + v_{q+1}^b}, & t_{q+1} - v_{q+1}^a \leq h_i < t_{q+1} + v_{q+1}^b \\ 0, & \text{elsewhere,} \end{cases}$$

and $\forall q \{0 < (v_q^a + v_q^b) < 1; \sum_q g_{qi} = 1\}$.

Trapezoidal coding is a partly linearized version of the semi-discrete coding. The fuzzy area around each knot is approximated by a linear function. Outside the overlapping fuzzy area the coding function is uniform and equal to either one or zero. The resulting function is visualized in Chapter 5. This type of coding is somewhere between first and second order B-splines because, if $(v_q^a + v_q^b) \to 0$, the trapezoidal coding converges to the first order B-spline (i.e. an ordinary step-function), if $(v_q^a + v_q^b) \to (t_{q+1} - t_q)$, the trapezoidal coding converges to piecewise linear coding and if $v_q^b \to t_{q+1}$ and $v_{q+1}^a \to t_q$, trapezoidal coding converges to the second order B-spline. The weak point, however, is again the arbitrary choice of the size of the fuzzy area around a knot. In most cases it is difficult to imagine reasonable if analytical, arguments for selecting non-arbitrary values for v_q^a and v_q^b $(q = 2, \ldots, r+1)$.

There exist some versions of this type of coding, which have positive uniform parts that are not necessarily equal to one and/or where the overlap can involve more than two coding functions. Le Foll (1979) proposes constant linear weights equal to 0.5 for each function, together with an overlap on three contiguous intervals. Another possibility is to interrupt the linear piece by one or more uniform parts ($\neq 1$). Anyway, such suggestions need a conceptual or theoretical basis, otherwise they only will enhance the arbitrariness of the whole procedure. In general as the overlap and the number of additional linear pieces increases, the global regularity of the transformation of the variable will increase. See the next two sections.

2.4. Coding by second order B-splines

This coding also sails under many colours: hat function, chapeau function, piecewise linear function and linear B-spline. First suggested as a way of coding data in homogeneity analysis by Bordet (1973), the piecewise linear function is well known in approximation theory and often mentioned in the context of regression splines, see Smith (1979).

The second order B-spline is defined by

$$g_{qi} = \begin{cases} \dfrac{h_i - t_q}{t_{q+1} - t_q}, & t_q \leq h_i < t_{q+1} \\ \dfrac{t_{q+2} - h_i}{t_{q+2} - t_{q+1}} & t_{q+1} \leq h_i < t_{q+2} \\ 0, & \text{elsewhere.} \end{cases}$$

Each coding function consists of two linear pieces defined on two adjacent intervals; each linear piece corresponds exactly with one interval. Because two coding functions overlap each other exactly in one interval, the weight of a datapoint is always spread over two linear functions. At a knot the functional value of one of them is, however, equal to zero. In Figure 2.1 we plotted the second order B-spline representation of the data of Table 2.1.

The piecewise linear coding adjusts to the local smoothness demand without any arbitrary constants. Different data values within an interval are not treated as being equal, as is predominantly the case in the preceding coding functions. With four knots we need four coding functions, all non-negative and the coding vectors are nearly orthogonal (see Table 2.2).

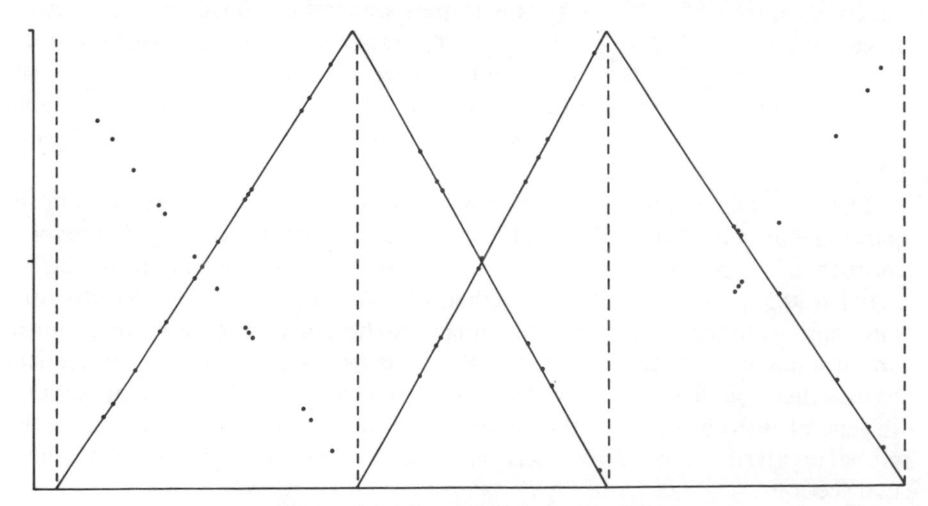

Figure 2.1. The second order B-spline

Table 2.1. Public spending in the Netherlands in the period 1951 until 1981, coded by crisp coding (A), semi-discrete coding (B) and trapezoidal coding (C)

	National public spending		*A*			*B*			*C*	
$t_1 = 30$										
	31.8	1	0	0	1	0	0	1	0	0
	32.4	1	0	0	1	0	0	1	0	0
	34.6	1	0	0	1	0	0	1	0	0
	33.3	1	0	0	1	0	0	1	0	0
	34.4	1	0	0	1	0	0	1	0	0
	34.4	1	0	0	1	0	0	1	0	0
	36.9	1	0	0	1	0	0	1	0	0
	38.1	1	0	0	1	0	0	1	0	0
	35.9	1	0	0	1	0	0	1	0	0
	36.2	1	0	0	1	0	0	1	0	0
	38.2	1	0	0	1	0	0	1	0	0
	38.4	1	0	0	1	0	0	1	0	0
	40.6	1	0	0	0.5	0.5	0	0.85	0.15	0
	40.9	1	0	0	0.5	0.5	0	0.77	0.23	0
	41.9	1	0	0	0.5	0.5	0	0.53	0.47	0
$t_2 = 43$										
	44.0	0	1	0	0	1	0	0	1	0
	45.8	0	1	0	0	1	0	0	1	0
	46.6	0	1	0	0	1	0	0	1	0
	46.4	0	1	0	0	1	0	0	1	0
	48.2	0	1	0	0	1	0	0	1	0
	50.5	0	1	0	0	1	0	0	1	0
	51.1	0	1	0	0	1	0	0	1	0
	51.5	0	1	0	0	1	0	0	1	0
	53.6	0	1	0	0	0.1	0.9	0	0.7	0.3
$t_3 = 54$										
	59.9	0	0	1	0	0	1	0	0	1
	59.8	0	0	1	0	0	1	0	0	1
	59.6	0	0	1	0	0	1	0	0	1
	61.6	0	0	1	0	0	1	0	0	1
	64.1	0	0	1	0	0	1	0	0	1
	65.4	0	0	1	0	0	1	0	0	1
	66.1	0	0	1	0	0	1	0	0	1
$t_4 = 67$										

According to De Boor's definition of B-splines, which is the definition that we use here, the coding function defined on the extreme left-hand and right-hand interval, i.e. the intervals between an interior and an exterior knot, $]-\infty, t_1[$ and $[t_u, +\infty[$, is truncated at the exterior knot. Outside the exterior knots all B-splines are equal to zero. In the French literature sometimes another convention is used, cf. Lafaye de Micheaux (1978), Le Foll (1979) and Martin (1980): outside the exterior knots the extreme coding functions are equal to one

Table 2.2. Public spending in the Netherlands in the period from 1951 until 1981 coded by first degree B-splines (D), semi-exponential coding (E) and by second degree B-splines (F)

	National public spending	*D*		*E*		*F*		
$t_1=30$								
	31.8	86	13	64	36	74	25	1
	32.4	82	18	62	38	66	32	2
	34.6	65	35	54	46	42	51	7
	33.3	75	25	59	41	56	41	3
	34.4	66	34	54	46	44	50	6
	34.4	66	34	54	46	44	50	6
	36.9	47	53	43	57	22	63	15
	38.1	38	62	36	64	14	65	21
	35.9	55	45	48	52	30	59	11
	36.2	52	48	46	54	27	60	13
	38.2	37	63	35	65	14	65	21
	38.4	35	65	34	66	13	65	22
	40.6	18	82	20	80	3	61	36
	40.9	16	84	17	83	3	59	38
	41.9	8	92	10	90	1	54	45
$t_2=43$								
	44.0	91	9	91	9	38	62	0
	45.8	75	25	75	25	25	72	3
	46.6	67	33	67	33	21	74	5
	46.4	69	31	69	31	22	74	4
	48.2	53	47	53	47	13	77	10
	50.5	32	68	32	68	5	74	21
	51.1	26	74	26	74	3	72	25
	51.5	23	77	23	77	2	70	28
	53.6	4	96	4	96	0	57	43
$t_3=54$								
	59.9	55	45	58	42	16	63	21
	59.8	55	45	59	41	17	63	20
	59.6	57	43	60	40	18	64	19
	61.6	42	58	50	50	9	57	34
	64.1	22	78	40	60	3	37	60
	65.4	12	88	35	65	1	22	77
	66.1	7	93	33	67	0	13	87
$t_4=67$								

instead of zero. This difference is irrelevant for finite coding, where we do not expect any data values outside the exterior knots. Le Foll (1979) interprets piecewise linear coding as the combination of crisp coding and dédoublement. The dédoublement of a real valued variable is equal to piecewise linear coding with only two exterior knots. This will produce a linear transformation of a variable (see Sections 3 and 4).

2.5. Coding by semi-exponential functions

Gallego (1980, p. 10) proposes a different variant of coding, that is piecewise linear within the interior knots and piecewise exponential on the extreme intervals. His definition, however, does not fit the analytical and graphical properties Gallego assigns to this coding. The following definition fits Gallego's graphical and verbal description and henceforth we call it the piecewise semi-exponential coding.

The definition of piecewise semi-exponential coding:

$$g_{1i} = \begin{cases} 1 - e^{(h_i - t_2)/(t_3 - t_2)}, & h_i \leq t_2 \\ 0, & \text{elsewhere,} \end{cases}$$

$$g_{2i} = \begin{cases} e^{(h_i - t_2)/(t_3 - t_2)}, & h_i \leq t_2 \\ (t_3 - h_i)/(t_3 - t_2), & t_2 < h_i \leq t_3 \\ 0, & \text{elsewhere,} \end{cases}$$

.
.
.

$$g_{qi} = \begin{cases} (h_i - t_{q-1})/(t_q - t_{q-1}), & t_{q-1} < h_i \leq t_q \\ (t_{q+1} - h_i)/(t_{q+1} - t_q), & t_q < h_i \leq t_{q+1} \\ 0, & \text{elsewhere,} \end{cases}$$

.
.
.

$$g_{u-1,i} = \begin{cases} e^{(t_{u-1} - h_i)/(t_{u-1} - t_{u-2})}, & h_i > t_{u-1} \\ (h_i - t_{u-2})/(t_{u-1} - t_{u-2}), & t_{u-2} < h_i \leq t_{u-1} \\ 0, & \text{elsewhere,} \end{cases}$$

$$g_{ui} = \begin{cases} 1 - e^{(t_{u-1} - h_i)/(t_{u-1} - t_{u-2})}, & h_i > t_{u-1} \\ 0, & \text{elsewhere.} \end{cases}$$

For $h_i \to \pm\infty$ the coding of the first and last interval goes to 1, while the partial derivatives in t_2 and in t_{u-1} coincide with the linear function on that interval. The motivation for these half-open extreme intervals is, that one wants an elegant coding of data-values outside $[t_2, t_{u-1}]$, which permits the reconstruction of all data-values outside this interval, such that no arbitrary definition of exterior knots t_1 and t_u is needed. The practical advantage could be that this is a more natural way of dealing with nutters (= outliers). According to Gallego this coding is most appropriate in case the data are uniformly distributed over intervals. The

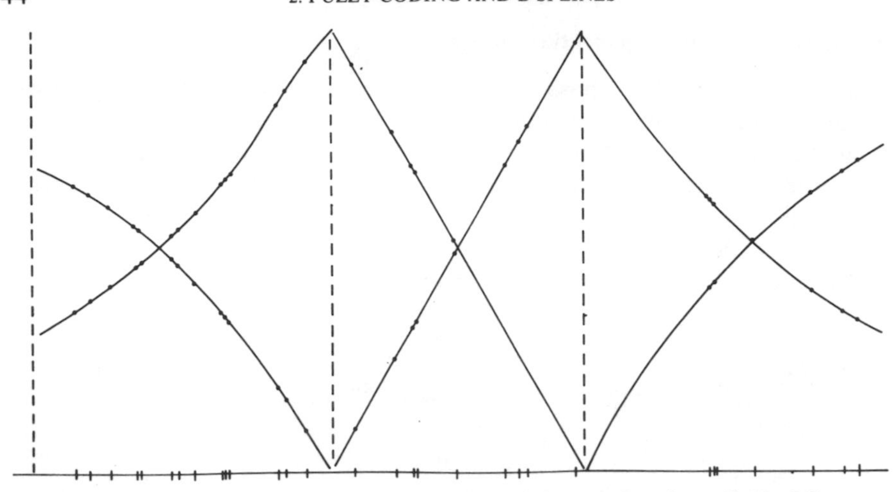

Figure 2.2. The semi-exponential coding of the test data from Table 2.2

practical, or data-analytical, relevance and the analytical validity of semi-exponential coding remain to be established. Additionally this type of coding seems to be not well conditioned, see Section 3 and Table 2.4 (E).

The coding by piecewise linear functions, and by first order B-splines in particular, has several desirable analytical global smoothing properties that are marginally discussed in Section 3 of this chapter. For an indepth study, one should consult De Boor (1978) or Schumaker (1981).

2.6. The coding by third order B-splines

The definition of the third order B-spline:

$$g_{qi} = \begin{cases} (h_i - t_{q-1})^2 (t_{q+1} - t_{q-1})(t_q - t_{q-1}), & t_{q-1} \leq h_i < t_q \\ (h_i - t_{q-1})(t_{q+1} - h_i)(t_{q+1} - t_{q-1})(t_{q+1} - t_q)^{-1} & \\ \quad + (h_i - t_q)(t_{q+2} - h_i)(t_{q+2} - t_q)(t_{q+1} - t_q)^{-1}, & t_q \leq h_i < t_{q+1} \\ (t_{q+2} - h_i)^2 (t_{q+2} - t_q)(t_{q+2} - t_{q+1})^{-1}, & t_{q+1} \leq h_i < t_{q+2} \\ \quad 0, & \text{elsewhere.} \end{cases}$$

This coding consists of quadratic, bell-shaped, functions that are non-zero on exactly three subsequent intervals. Every coding function overlaps the next function in exactly two intervals, which, in case of our test data with a knot sequence of four knots, leads to five coding functions. At each exterior breakpoint we need a knot multiplicity equal to two (i.e. two knots at the same location) because exactly two coding functions are positive at each location. In Figure 2.3 this coding function is plotted against the raw data with the same legend as in

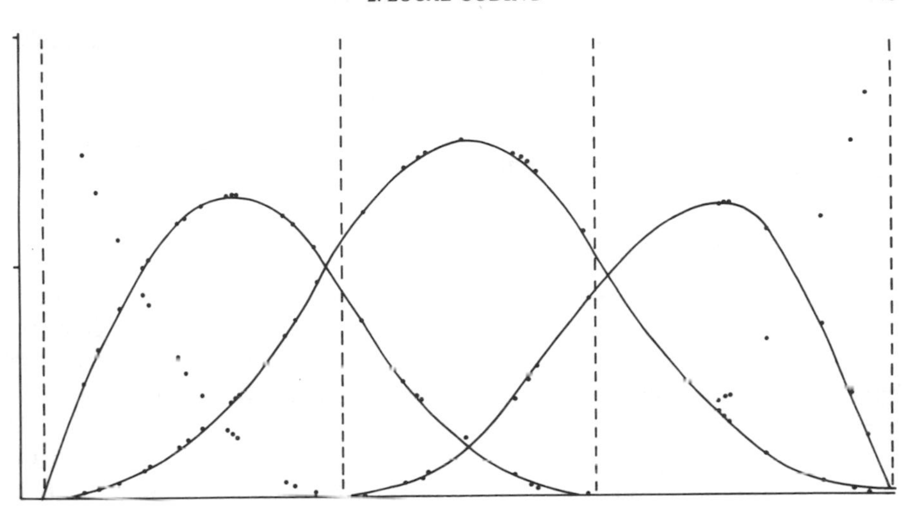

Figure 2.3. The third order B-spline

Figures 2.1 and 2.2. While inspecting the plot, the mass of overlapping functions is difficult to entangle and it is a euphemism to call this a simple coding system. The coding matrix consists of non-negative numbers, is nearly orthogonal and has five columns. See Table 2.2.

2.7. Coding by piecewise bivariate functions

An obvious generalization of the low-dimensional univariate coding function is the bivariate coding function. The paired intervals of two variables are coded by one bivariate coding function in R^2. See Figure 2.4.

The global bivariate coding surface consisting of overlapping piecewise bivariate functions is related to the scalar-product spaces as discussed in Chapters 4 and 6, see also Besse and Ramsay (1986). The latter, however, are

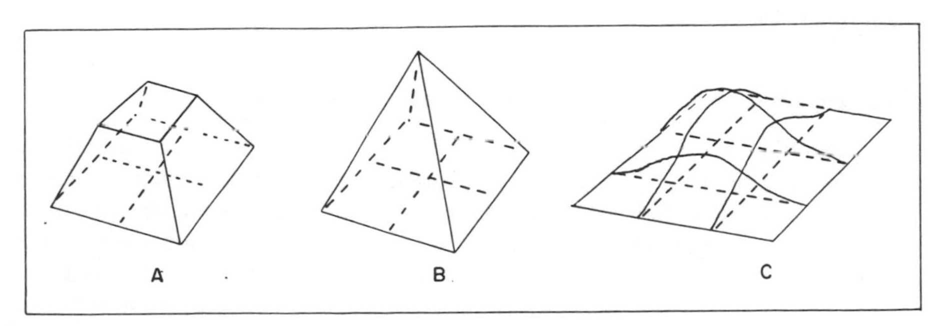

Figure 2.4. Bivariate coding functions: bivariate trapezoidal function (A), Tensor product B-splines of degree 1 (B) and of degree 2 (C)

defined on pairs of variables and not on pairs of segments of different variables.

Bivariate coding of subsets is often mentioned by other authors but the idea as such is hardly extended any further in data analysis because of reasons of feasibility. In spline theory such bivariate functions are known as tensor product B-splines (cf. Schumaker, 1981, Ch. 12).

3. GLOBAL TRANSFORMATION

Each type of coding defines a class of transformations of the data. In this section we are interested in what way such a transformation function is generated by the coding systems from Section 2, in order to understand the smoothness of the global transformation. Lafaye de Micheaux (1978) shows that nearly all functions for fuzzy coding from Section 2 form a basis for functions in L^2, the space of square integrable continuous functions; see also Chapters 4 and 6 and Besse and Ramsay (1986). For an introductory review of all essential ideas, see Ramsay (1982).

3.1. The orthogonality of fuzzy coding systems

When verifying the orthogonality of the scalar-product matrices of all types of fuzzy coding discussed so far we see that the near orthogonality, defined in Section 2, holds for all fuzzy codes except for the semi-exponential coding. See Table 2.3. The number of overlapping coding functions determines the number of sub-diagonals in the matrix of scalar-products D. In case of crisp coding D is equal to a diagonal matrix (4A). The semi-discrete (4B), trapezoidal (4C), linear (4D) and semi-exponential coding (4E) have all tri-diagonal scalar-product matrices. Quadratic coding (4F) has a penta-diagonal scalar-product matrix. The bandwidth of such a matrix is equal to the number of non-zero (sub)diagonals. In case of basis of B-splines with degree v the bandwidth of the scalar-product matrix is equal to $2v+1$. Note that all matrices are symmetric and non-negative. Such scalar-product matrix is called the matrix of coefficients in approximation theory.

3.2. The smoothness and shape of fuzzy transformation functions

The smoothness of Φ is determined by the degree of the functional pieces and their behaviour at the knots. The sum of two functions of degree v is equal to another function of degree v. This means that between two arbitrary points on $[ab]$, Φ has a degree that is equal to the degree of the corresponding functional pieces. For example in case of crisp coding the degree of Φ is equal to zero everywhere and Φ is not connected in the knots. If Φ is based on piecewise linear functions Φ will be linear within all intervals and will be connected in the knots. With a basis of quadratic B-splines Φ is quadratic in all intervals and once differentiable in the

Table 2.3. The scalar product for crisp coding (A), semi-discrete coding (B), trapezoidal coding (C), linear coding (D), semi-exponential coding (E) and quadratic coding (F)

(A)	15				
		9			
			7		
(B)	13.5	0.75			
	0.75	9.6	0.09		
		0.09	7.9		
(C)	13.59	0.55			
	0.55	8.78	0.21		
		0.21	7.9		
(D)	4.52	2.94			
	2.94	7.41	1.58		
		1.58	4.20	1.33	
			1.33	3.17	
(E)	3.17	3.34			
	3.34	8.19	1.58		
		1.58	4.71	1.64	
			1.64	1.99	
(F)	2.13	2.05	0.34		
	2.05	4.79	2.38	0.07	
	0.36	2.38	5.25	1.33	0.16
		0.07	1.33	2.11	1.08
			0.16	1.08	1.94

knots; i.e. at an interior knot Φ consists of the sum of two quadratic functions and the first derivative thus does exist. If Φ is a polynomial spline function the smoothness and degree of Φ are exactly defined by the degree of the B-spline basis. For fuzzy coding in general Φ can be piecewise of different degree, Φ can vary in the amount of overlap and the functional pieces are not necessarily defined on the exact intervals defined by knot sequence $\{t\}$.

As to how the choice of coding functions affects the shape of the global transformation function and which role is played by the weights $\{\alpha\}$, we use some sets of artificial weights that are all linear in the knot sequence $\{t\}$. The data from Table 2.1 are coded by the three different coding systems, discussed in Section 2. The transformation functions corresponding with semi-discrete and trapezoidal coding are treated in Chapter 5.

The linear and semi-exponential coding are weighted by $\{\alpha\} = \{1, 1.33, 1.66, 2\}$

and the quadratic coding by $\{\alpha\}=\{1, 1.25, 1.5, 1.75, 2\}$. The number of coefficients is always equal to the dimensionality of Φ. We inspect the shape and smoothness of the global transformation by plotting Φ against the data. See Figures 2.5 and 2.6.

The linear coding, or first degree B-spline, builds a piecewise linear transformation function that is connected in the knots, because

$$\Phi(t_q)=\alpha_{q-1}G_{q-1}(t_q)+\alpha_q G_q(t_q),$$

and

$$\Phi(t_q)=\alpha_q G_q(t_q)+\alpha_{q+1}G_{q+1}(t_q),$$

and

$$\alpha_{q-1}G_{q-1}(t_q)=\alpha_{q+2}G_{q+2}(t_q)=0.$$

Because the sum of two linear functions within an interval is another linear function we know that Φ is piecewise linear. Note that it is necessary that the overlap is defined in terms of intervals (just one interval in this case).

The shape of the global transformation function of the semi-exponential coding is smoother than linear coding, and less regular than quadratic coding. See Figure 2.5(B).

Because the sum of three overlapping quadratic functions is equal to another quadratic function, we know that the transformation function corresponding to quadratic coding is piecewise quadratic. Φ is also quadratic at the knots because

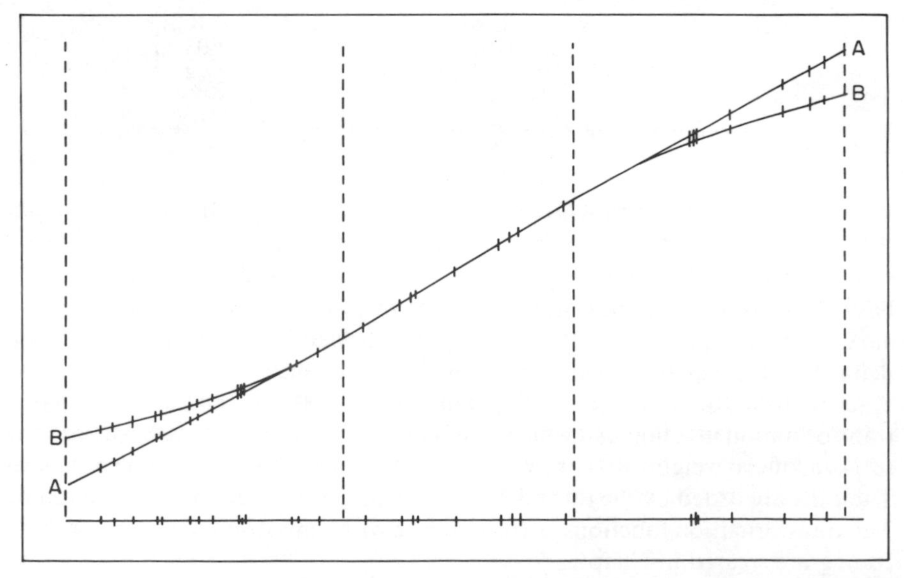

Figure 2.5. The piecewise linear (A) and the semi-exponential (B) transformation function

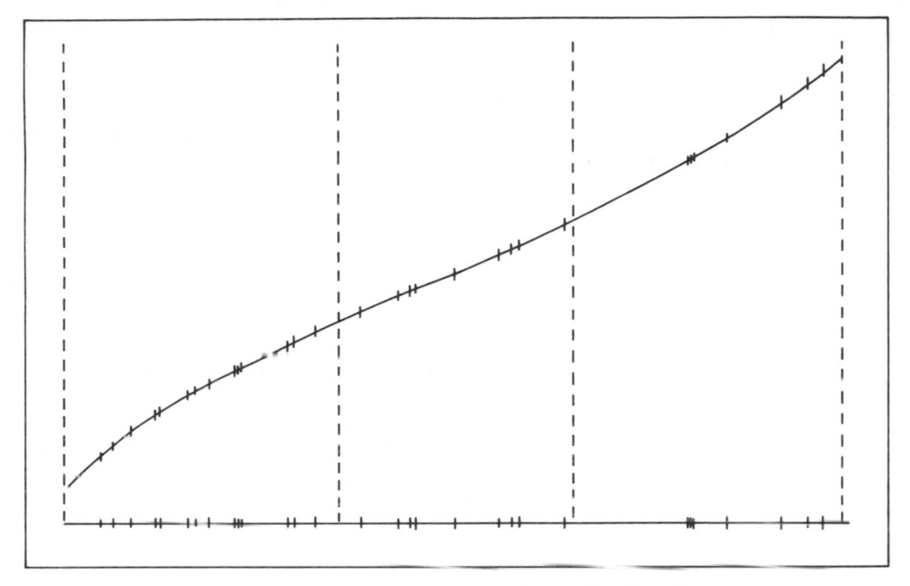

Figure 2.6 The quadratic transformation function

it is equal to the sum of two quadratic functions. Therefore the function is smooth up to the first derivative. Note that the increase of smoothness of the six coding systems, from Section 2, is defined by the amount of overlap per interval and the degree of the coding function.

4. GLOBAL APPROXIMATION

4.1. Introduction

We interpret Φ (h) as an approximation of a model vector, that is defined by some model or technique. The model vector is defined by the vector x. Chapter 3 is devoted to the embedding of Φ in homogeneity analysis, where x is equal to the vector of observation scores on a dimension. In this section we accept the availability of x without further reference.

In the terminology of functional theory $\Phi(h)$ is the approximation of a function (i.e. the model vector) based on a representation (=fuzzy coding) on some finite amount of data on that function (= the data vector). Schumaker (1981) calls such a problem a data-fitting problem. The data vector is always subjected to error or noise and moreover, does not determine the model vector uniquely. In such a case the least squares fit criterion combined with a well conditioned basis is an attractive norm for approximation, cf. De Boor (1978, p. 249).

Before we proceed it may be useful to connect back to the last section: we observed that the coefficients $\{\alpha\}$ are the only determining parameters for Φ.

These coefficients are selected in such a way that Φ approximates the model vector x as well as possible according to the least squares fit criterion. This approximation is based on the piecewise coding of the data vector. The shape of Φ is hence predetermined by three factors: the functional type of coding, the knot sequence $\{t\}$ and the least squares fit of Φ to x.

In this section we will discuss two determining factors on Φ: the optimal properties of Φ and the *proper* choice of knot sequence $\{t\}$. The least squares approximation of x by $\Phi(h)$ is numerically a rather routine affair, i.e. solving a set of normal equations to be found in any textbook on numerical analysis; see also Chapter 3.

4.2. Optimality properties

An introduction to the general idea used in this section is provided by Ramsay (1982). We have to make the distinction between fuzzy coding in general and B-splines in particular. The optimality properties of the global transformation function based on fuzzy coding in general are not widely discussed.

4.2.1. *Fuzzy coding in general*

As far as the most primitive form (=crisp coding) is concerned, Dauxois and Pousse (1976) prove that crisp transformation is dense in L^2, the space of square integrable continuous functions, if the number of knots goes to infinity. Lafaye de Micheaux (1978, Ch 2.2) extends this proof to trapezoidal and piecewise linear functions. He recommends the use of a basis of B-splines, if the model vector to be approximated has some regularity properties (Lafaye de Micheaux, 1978, p. 76). Both Martin (1980) and Lafaye de Micheaux (1978) prove that all coding functions from Section 2, except for semi-exponential coding, are convergent in both discretization and sample size (Lafaye de Micheaux, 1978, section 3.1). Martin (1980) expects a faster rate of convergence in general for both aspects if a basis of splines rather than the more general basis of fuzzy codes is used. Gallego (1980) remarks that looking for that piecewise coding function with the fastest rate of convergence, is the main objective for further research on piecewise coding functions.

4.2.2. *Spline functions*

The optimality of the polynomial spline function in terms of approximation is widely covered in literature, cf. De Boor (1978), Schumaker (1981) and many others. The general requirements for approximating functions are smoothness, computational ease and membership of a large class of approximating functions, such that arbitrary smooth functions can be approximated. Polynomials seem to fulfil these requirements, but they have one drawback: the shape of a polynomial is rigid. In a low degree approximation the shape of a polynomial is too rigidly defined by very few points and higher degree polynomial approximations oscillate wildly in general. This is called the inflexibility of the class of approximating polynomials. This inflexibility becomes manifest mainly in polynomials of order ≥ 3 on large intervals. So we need polynomials of a low degree defined on small intervals. Piecewise polynomials of a low degree would be sufficient, but they are not necessarily connected in the intervals and thus not necessarily smooth between intervals, only within intervals. Schumaker gives the following list of attractive features of polynomial splines:

—Polynomial spline spaces are finite-dimensional linear spaces with very convenient bases;
—Polynomial splines are relatively smooth functions;
—Polynomial splines are easy to store, manipulate and evaluate on a computer;
—The derivatives and antiderivatives of polynomial splines are again polynomial splines, whose expansions can be found on a computer;
—The number of zeros of a polynomial spline of order v cannot exceed $v-1$ and thus they cannot oscillate too wildly;
—Various matrices arising naturally in the use of splines in approximation and numerical analysis, are always non-singular and have strong sign regularity properties;
—The sign structure and shape of a polynomial spline can be related to the sign structure of its coefficients;
—Low order splines are very flexible and do not exhibit the oscillations usually associated with polynomials;
—Every continuous function on the interval (a, b) can be approximated arbitrarily well by polynomial splines with the order $v+1$ fixed, provided a sufficient number of knots are allowed;
—Precise rates of convergence can be given for approximation of smooth functions by splines—not only are the functions themselves approximated to high order, but their derivatives are simultaneously approximated as well.

The latter convergence properties are related to the convergence in discretization, discussed in Section 1.

A spline, a word used first in a mathematical context by Schoenberg (1946a, 1946b), is an elastic strip of wood that has to go through a number of fixed points called ducks mainly used in shipbuilding. Euler and the Bernoulli brothers already found out that the centreline of the elastic rod is approximately given by the positions of the ducks. In a similar way a spline function is determined by the function values in the knots.

Spline theory, without using the word spline, goes back to Runge (1901), Eagle (1928), Quade and Collatz (1938) and others. The spline function seems to have been rediscovered around 1960. For a survey see Schumaker (1981). A complete bibliography on spline functions up to 1973 is given by Van Rooij and Schurer (1974).

In data analysis splines are generally used as a non-parametric function estimation technique. The main areas of application are (see also Wegman and Wright, 1983):

—*Non-parametric regression* (i.e. the smoothing spline of Schoenberg and Reinsch) See Wahba (1978).
—*Regression splines*. Smith (1979), Wold (1974), Poirier (1973, 1976).
—*Non-parametric density estimation*. Boneva, Kendall and Stefanov (1971), Wahba (1971), Good and Gaskins (1971).
—*Isotone regression*. De Vore (1977), Wright and Wegman (1980), Winsberg and Ramsay (1980, 1981, 1983).
—*Non-linear multivariate analysis*. Lafaye de Micheaux (1978), Gifi (1981a, 1988), Van Rijckevorsel (1982, 1987), Besse and Ramsay (1986).
—*Time series analysis*. Cogburn and Davis (1974).

4.3. Knot placement

4.3.1. *Introduction*

The analytical aspects of knot placement in this section are defined in relation to polynomial spline functions. However, we would like to invite the reader to apply these results, albeit on an intuitive level, to the more general forms of fuzzy coding, mainly because the corresponding global functions are (also) dense in L^2.

There exist several items to consider in a discussion on knot placement.

(1) The dimension of a spline function is a simple function of the degree and the number of interior knots, see Section 3. If the dimensionality is fixed, the smaller the number of knots, the higher the degree and vice versa.
(2) We can always change the location of the knots; this does not affect the dimensionality of the polynomial spline function.
(3) If we use a (quasi) uniform knot placement, the location problem does not exist and the optimal knot placement is reduced to the determination of the optimal number of knots.
(4) The optimal knot placement is sometimes regarded as the choice of model and not as a free parameter to be optimized, i.e. in the same way as the degree of the approximating function or the dimensionality in a component analysis is chosen.
(5) The knot placement can influence the fit of a subsequent model. In so far as this fit is determined by the sample size and the marginal distribution over intervals, there will be a direct relationship between knot placement and fit.
(6) Optimality can be defined with regard to approximation or to data analysis. These aims do not necessarily coincide.

4.3.2. *Approximation and knot placement*

There exist some saturation and convergence theorems with lower- and upperbounds of approximation for (quasi) uniform and free knot sequences if the number of knots goes to infinity, see Schumaker (1981, Chs 6 and 7). Important conclusions with respect to our goal are that, if the function to be approximated is smooth enough while the number of knots goes to infinity, the uniform knot sequence is never inferior to a free knot sequence. For less smooth functions the rate of convergence can be considerably better using free knots instead of uniform knots. *Ergo*, there is theoretically something to be gained by optimal free knot placement.

Apart from the analytical results, it is obvious that with an increasing dimensionality, caused by an increase of degree, number of knots or both, there exists the realistic danger of overfitting, because of the flexibility of the polynomial spline function. This provides us with a reason to prefer low-dimensional spline functions, regardless of the knot placement. Because we are mainly interested in global functions of a degree equal to or less than three, this virtually amounts to using as few knots as possible.

Knot optimization can only be performed at considerable computational cost and for a certain fit value there may exist many *best* knot placements. Only if the function to be approximated has distinct and explicitly known discontinuities and there exists a need for a low-dimensional approximation, search methods for knot optimization should be considered, cf. De Boor (1978, Ch. XII). If the distance between the polynomial spline function and the function to be approximated cannot be evaluated properly because of a lack of functional information of the latter, optimal knot placement becomes irrelevant.

Concluding we observe that knot optimization is mainly useful as a free parameter, if we know the function to be approximated well enough and if we know it to be discontinuous. And even then, if we would accept a greater dimensionality, a uniform knot placement with a generous number of knots would do just as well, cf. De Boor (1978, p. 272).

4.3.3. *Smoothness, least squares estimation and knot placement*

It is obvious that with a knot at every data-value we will obtain maximal smoothness by interpolation, which contradicts the least squares criterion. The penalized least squares smoothing spline of Schoenberg and Reinsch actually uses this situation, by weighting the amount of smoothing with a knot at every data-value against the least squares fit. This approach has however a limited use, because the precision of the data must be well known, the function to be approximated must be equidistant and the number of data values large, cf. Wold (1974).

Knowing this, it is not surprising that several authors concentrating on least squares regression applications with limited information on the function to be fitted, prefer to see (optimal) knot placement as a model choice similar to the choice of the type of fuzzy coding; cf. Wold (1974), Smith (1979), Poirier (1973, 1976), Winsberg and Ramsay (1981). Local knot placement improvement schemes on the other hand, defined within a particular technique are developed by Hudson (1966), Park (1978), Agarwal and Studden (1978) and Gallego (1980). A more general approach to optimal knot placement is given by De Boor (1978).

The fit of noisy data by means of least squares splines needs generally low-dimensional spline functions with few knots and whether a local improvement scheme for knot placement will work, will largely depend on the size and precision of the data. Do not forget: noisy data are not precise!

4.3.4. *Relevance for data analysis*

- Knot optimization has limited value for noisy data.
- Always use a uniform knot sequence unless the nature of the data compels you to act differently.
- Always use low-dimensional global transformation functions.
- Be careful with small datasets.
- Always use as few knots as possible.
- Always have at least five data values between two sequential knots, cf. Wold (1974).
- Always use any information whatsoever on discontinuities for a knot placement; particularly with respect to expected inflection points and extrema.

Component and Correspondence Analysis
Edited by J.L.A. van Rijckevorsel and J. de Leeuw

Chapter 3

Beyond Homogeneity Analysis

Jan de Leeuw
Department of Psychology and Mathematics
UCLA, Los Angeles, USA
and
Jan L.A. van Rijckevorsel
Department of Statistics, TNO NIPG, Leiden, The Netherlands

1. INTRODUCTION

In Gifi (1981a, 1988) a large number of multivariate analysis methods are organized in a single general framework. The key method in this system is *homogeneity analysis*, also known as *multiple correspondence analysis* (cf. Chapter 1). The Gifi system is inspired by ideas from multidimensional scaling, in particular by the central role of Euclidean distance in the representation of complex multivariate data. The basic data we want to represent geometrically are categorizations of n objects by m variables. Although the assumption that the variables are discrete and assume only a finite number of values is not essential, and can even be made without any practical loss of generality, it is true that in the current versions of homogeneity analysis categorical variables with a small number of categories play a central role. Variables with a large number of possible values, or even 'continuous' variables, can be incorporated in theory, but the implementations of the techniques more or less expect a small number of categories. If the number of categories is very large, say close to the number of objects that are classified, then homogeneity analysis as currently implemented (Gifi, 1981b) does not work very well. It will tend to produce unsatisfactory and highly unstable solutions, in which 'chance capitalization' is a major source of variation (cf. also Chapter 2).

There have been various attempts to make the solutions more stable by imposing restrictions that reflect, in some sense, the prior information we have about the variables. In De Leeuw (1984a) these restrictions are classified into *rank-restrictions*, *cone-restrictions* and *additivity-restrictions*. Imposing restrictions decreases the number of free parameters. This means, roughly, that there are more data values per parameter, which can consequently be determined in a more stable manner. Rank-restrictions and cone-restrictions make it more easy to deal with variables having a large number of categories, but in several respects their treatment remains somewhat unsatisfactory. In many multidimensional scaling programs there are options for transformation of the variables that are 'smooth' or otherwise 'continuous'. There is no such possibility in the current homogeneity analysis programmes. In this chapter we shall try to extend the basic geometry of homogeneity analysis in such a way that continuous variables fit in more easily using the coding systems as discussed in Chapter 2. A fundamental role in this extension is played by the 'B-spline basis' and its 'fuzzy' generalizations, which is introduced here in a purely geometrical way, that is mainly due to van Rijckevorsel (1987). This additionally indicates more clearly how homogeneity analysis generalizes the various forms of non-metric principal component analysis (cf. Chapter 1). Combination of the various options creates a very flexible new type of homogeneity analysis. It is highly unlikely that all possible combinations will be equally important in practice, in fact we suspect that some of the less restricted forms will again tend to produce very unstable or even 'trivial' solutions. Nevertheless it is satisfactory from a theoretical point of view to show exactly what the choices are that one has to make, even if some of the possible choices may be quite unwise in practical situations.

2. SIMPLE HOMOGENEITY ANALYSIS

We start with a brief recapitalization of the technique of homogeneity analysis introduced in Chapter 1, without any of the frills discussed by Gifi (1981a, 1988) or de Leeuw (1984a, 1984b). The data are m variables on n objects, i.e. there are m functions defined on a common domain $\{1, 2, \ldots, n\}$. We suppose that the range of function j has k_j elements, and we code function j by using the $n \times k_j$ *indicator matrix* G_j. Matrix G_j is binary, it has exactly one element equal to one in each row, indicating into which element of the range the object corresponding to this row is mapped. Thus the rows of G_j add up to one, and the matrix $D_j = G_j'G_j$ is diagonal, and contains the univariate marginals. $G_{jl} = G_j'G_l$ is the cross-table of variables j and l and contains the bivariate marginals. This notation is illustrated in detail in Chapter 1.

The purpose of homogeneity analysis is to map both objects and variables into low-dimensional Euclidean space R^p (where p is *dimensionality*, chosen by the user). We want to do this in such a way that both objects and categories of the variables are represented as points, and in such a way that an object is relatively close to a category it is in, and relatively far from the categories it is not in. Of course this implies, by the triangle inequality, that objects mostly scoring in the same categories tend to be close, while categories sharing mostly the same objects tend to be close too. The extent to which a particular representation X of the objects and particular representations Y_j of the categories, satisfy the desiderata of homogeneity analysis is measured by a least squares loss function. This is defined as

$$\sigma(X; Y_1, \ldots, Y_m) = \sum_j \text{tr}(X - G_j Y_j)'(X - G_j Y_j). \tag{2.1}$$

In order to prevent certain obvious trivialities we require that the $n \times p$ matrix of *objects scores* X is normalized by $u'X = 0$ and $X'X = nI$. Here u is a vector with all elements equal to one, and I is the identity matrix. We do not normalize the m matrices of *category quantifications* Y_j, which are of order $k_j \times p$. Using (2.1) and the normalization conventions we can now give a more precise definition of homogeneity analysis. It is to choose a normalized X and $Y_1, \ldots, Y_m$ in such a way that (2.1) is minimized. For additional interpretations of the loss function, in terms of consistency discrimination and homogeneity, we refer to Gifi (1981a) and de Leeuw (1984a). In this Chapter we more or less ignore the algorithmic and statistical aspects of the homogeneity analysis techniques, and concentrate on the geometry on which the loss function is based.

3. PICTURES OF LOSS

In Table 3.1 we have presented a small example with ten objects and three variables. The objects are ten cars, the variables are price (in $1000), gas consumption (litres per 100 km, on the expressway) and weight (in 100 kg). The data are taken from Chapter 6, Table 6.1. In order to prevent possible misunderstandings we must emphasize that Table 3.1 in this chapter is not at all representative for data usually analysed with homogeneity analysis. In fact, in most practical applications of the technique, the number of objects and the

Table 3.1. Car data

	Price	*Gas*	*Weight*
Chevette	5.6	6.9	9.7
Dodge Colt	5.7	5.1	8.8
Plymouth Horizon	6.3	5.5	9.9
Fort Mustang	7.6	6.7	12.0
Pontiac Phoenix	8.6	6.9	12.1
Dodge Diplomat	9.4	10.2	15.5
Chevrolet Impala	10.1	7.5	16.9
Buick Regal	10.5	7.8	15.0
AMC Eagle	10.7	11.7	15.7
Oldsmobile 98	13.3	8.7	18.3

number of variables is much larger. Moreover in our small example all variables are numerical, which is also not typical for most homogeneity analysis applications.

The data in Table 3.1 cannot be used directly in homogeneity analysis. They must first be made discrete or categorical. This is done by grouping the values of the variables into discrete categories, which can, of course, be chosen in many different ways. One possible, fairly crude, categorization is given in Table 3.2. Observe that there are three cars with *profile* (1, 1, 1), and two cars with (2, 1, 2). Thus there are only seven different profiles for these ten cars, out of possible $3 \times 3 \times 4 = 36$ profiles. A finer discretization would give more possible profiles, more different actual profiles, and also more 'empty cells', i.e. more profiles that do not occur. The finest discretizations is the *ranking* given in Table 3.3. Here there are $10^3 = 1000$ possible profiles, of which only 10 are in use. Thus 99 per cent of the cells are empty. Observe that in constructing Table 3.3 from Table 3.1 we have arbitrarily broken a tie in variable 2 (Chevette and Pontiac Phoenix both score 6.9 in gas consumption).

Table 3.2. Car data, discrete

	Price	*Gas*	*Weight*
Chevette	1	1	1
Dodge Colt	1	1	1
Plymouth Horizon	1	1	1
Fort Mustang	2	1	2
Pontiac Phoenix	2	1	2
Dodge Diplomat	2	3	2
Chevrolet Impala	3	2	3
Buick Regal	3	2	2
AMC Eagle	3	3	2
Oldsmobile 98	4	2	3

Table 3.3. Car data, ranked

	Price	*Gas*	*Weight*
Chevette	1	4	2
Dodge Colt	2	1	1
Plymouth Horizon	3	2	2
Fort Mustang	4	3	4
Pontiac Phoenix	5	5	5
Dodge Diplomat	6	9	7
Chevrolet Impala	7	6	9
Buick Regal	8	7	6
AMC Eagle	9	10	8
Oldsmobile 98	10	8	10

Now suppose we choose object scores X in two dimensions, and category quantifications Y_j also in two dimensions. We have plotted the objects scores we have chosen as ten points in Figure 3.1. Also given in Figure 3.1 are the three points corresponding with the categories of variable 1, price. To make a picture of loss, for variable 1, we have connected all objects with the category point they belong to, according to variable 1. Loss-component 1 is simply the sum of squares of the line-lengths drawn in Figure 3.1. We can make a similar picture for variable 2, if we also choose Y_2. It is important to realize that we have chosen X

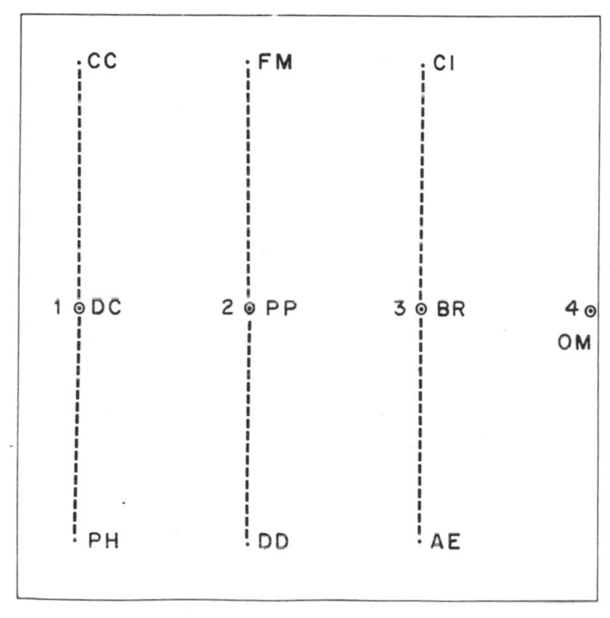

Figure 3.1. Loss variable 1, arbitary solution

and Y_1 completely arbitrary, and not by any optimality considerations. They are not, in any sense, the solutions given by homogeneity analysis. In fact they are merely candidates for the solutions, and it is the purpose of the technique to find better candidates. Another important point is that we can also make 'dual' pictures, in which we plot all Y_j as points together with a single object point. The loss 'due to object i' can now be represented by drawing lines from the object point to all category points it is in. Such plots, as well as the plot in Figure 3.1, are 'sub-plots' of a large plot which contains all object points and all category points, and which has a line for each element equal to one in each indicator matrix. This 'super-plot' will generally look somewhat messy, so it is better to present it in 'layers'. In Figure 3.2 we have presented the optimal solution computed by homogeneity analysis, i.e. the optimal object scores and the optimal quantifications of the categories of variable 1. It is clear that the line lengths are shorter for

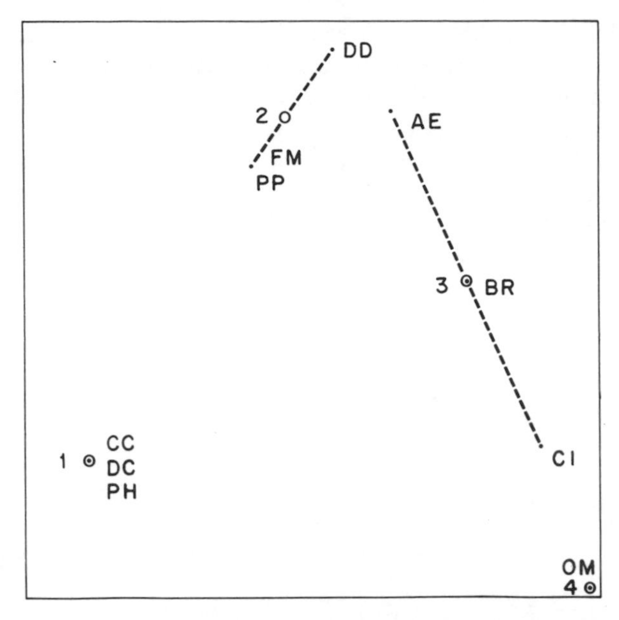

Figure 3.2. Loss variable 1, optional solution

the optimal solution. For other types of plots useful in homogeneity analysis we refer to Gifi (1981a, 1981b, 1988).

4. RANK RESTRICTIONS

In simple homogeneity analysis category quantifications can be anywhere in p-space. From equation (2.1) it follows that optimal category quantifications are centroids of objects points in the categories. This is illustrated in Figure 3.2. In fact in Figure 3.1 category quantifications of variable 1 are also optimal for the given object scores, only the object scores are very far from optimal in this case. Because of the *centroid-property* of optimal category quantifications it follows that their weighted average, with weights equal to the marginal frequencies, is the origin. This is the only restriction on the relative position of the quantifications of the categories within a variable. Now consider the situation in which variables have a range which is ordinal or even numerical. This constitutes a form of prior information which is not used by simple homogeneity analysis, and which consequently may get lost in the representation computed by homogeneity analysis. If we look at Figure 3.2 the categories of variable 1 are represented in the 'correct' order. This is true if we measure order along the horizontal axis, and even more clearly true if we measure order along the 'horseshoe' on which all objects lie. For variable 2, gas consumption, the situation is quite different, however. Only Dodge Diplomat and AMC Eagle are in category 3, which means that the optimal quantification of the category will be the midpoint of the line connecting DD and AE. Category 2 contains CI, OM and BR, and will be quantified close to CI. Category 1 will be between cluster CC, DC, PH and cluster PP, FM. Thus both on the horseshoe and on the line the categories will project in the order 1–3–2, which is contrary to our prior information. In this chapter we discuss geometrically inspired methods which both prevent the horseshoe and make it possible to impose our prior information.

A familiar way to get rid of the horseshoe is to do this by imposing rank-one restrictions (van Rijckevorsel, 1987). By this we mean that we require all category quantifications of a variable to be on a line through the origin of p-space, with each variable having its own line. In matrix notation this means that we require $Y_j = z_j a_j'$, i.e. the $k_j \times p$ matrix Y_j must be of rank one. In Chapter 1, in order to distinguish the various types of category quantifications that result from this idea the Y_j are called *multiple category quantifications*, while the z_j are called *single category quantifications*. The a_j are the loadings of variable j. We now minimize the loss function (2.1), with the provision that for some variables (but not necessarily for all) we use the restrictions $Y_j = z_j a_j'$. Variables for which the restrictions are imposed are called *single variables*, variables without restrictions are *multiple variables*. A program for homogeneity analysis with mixed multiple and single variables is discussed by Gifi (1982).

In order to study the geometry of single variables we expand the corresponding loss component first. This gives

$$\begin{aligned}\operatorname{tr}(X - G_j Y_j)'(X - G_j Y_j) &= \operatorname{tr}(X - G_j z_j a_j')'(X - G_j z_j a_j') \\ &= np - 2a_j' X' G_j z_j + (z_j' G_j' G_j z_j)(a_j' a_j). \qquad (4.1)\end{aligned}$$

Now let $q_j = G_j z_j$, and normalize z_j such that $u'q_j = 0$ and $q_j' q_j = n$. Such normalization is used merely for identification purposes, because z_j only occurs in the product $z_j a_j'$. Using the normalization we find

$$\operatorname{tr}(X - G_j Y_j)'(X - G_j Y_j) = n(p-1) + (q_j - X a_j)'(q_j - X a_j). \qquad (4.2)$$

This shows, in the first place, that single loss cannot possible be zero if p is larger than one. It is always at least $n(p-1)$. It is equal to $n(p-1)$ if all objects in a category project in the same point on the line through the origin and a_j. Or, to put it differently, if categories define parallel hyperplanes orthogonal through the line defining the variable. All objects in a category must be located in the hyperplane of the category. The elements of z_j are the signed distances to the origin of the category hyperplanes, i.e. the location of the projections on the line defining the variable. In the case of non-perfect fit the loss is simply the distance of each object point from its category hyperplane, or, more precisely, the squared distance. Figure 3.3 illustrates this for a particular choice of X, z_1, and a_1 in our small car example. Again no optimality considerations are used here, in fact we have not even paid attention to the appropriate normalizations. It is clear that rank-one

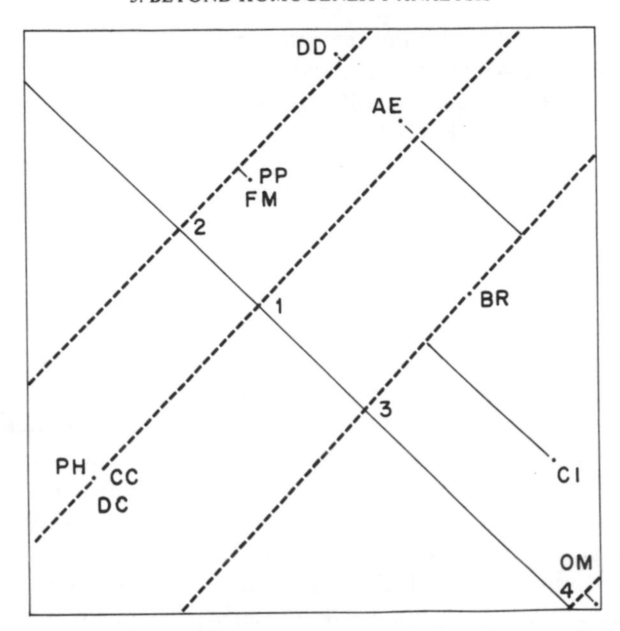

Figure 3.3. Single nominal loss, variable 1, arbitrary solution

restrictions will tend to make horseshoes impossible, or at least highly unlikely. It may not be clear yet how they can be used to impose ordinal or numerical prior information. Before we proceed to explaining this, remember that the use of single variables is related to performing a *principal component analysis* as in Chapter 1.

5. CONE RESTRICTIONS

Rank-one restrictions induce an order on the categories of the variable, even if we do not know the order beforehand. The induced order is given by the projections on the variable vector, or by the order of the category hyperplanes. In fact the category hyperplanes even introduce a single numerical scale for the categories of a variable, given in the vector z_j. Now the induced ordinal or numerical information may or may not correspond with our prior knowledge. We use *cone restrictions* if we impose the constraint that the induced order must be the same as our prior order, and the induced scale must be the same as our prior scale. Numerically these are restrictions on the elements of z_j. Either they must be in the 'correct' order, for *single ordinal* variables, or they must be equal to a given normalized vector, for *single numerical* variables. Observe that the type of a variable refers to the constraints we impose, it does not reflect some intrinsic property of the variable. We use the term 'cone restrictions' because the feasible choices for z_j form a polyhedral convex cone in k_j-space for ordinal variables, and a one-dimensional subspace, which is a sort of degenerate cone, for numerical variables. It is also possible, as is done in Chapters 4, 5 and 6, to formulate our restrictions in terms of $q_j = G_j z_j$, i.e. in n-space or in the scalar-product space of vectors q_j ($j = 1, \ldots, m$). No restrictions on z_j, defining *single nominal* variables, defines a k_j-dimensional subspace in n-space. Ordinal and numerical restrictions defines subcones and subspaces of this k_j-dimensional subspace.

If the z_j are completely given, by restrictions taken together with normalizations, then homogeneity analysis becomes identical with principal component analysis, cf. Chapter 1. This is, in a sense, one of the endpoints of the continuum of homogeneity analysis techniques. All variables are single numerical; the other endpoint has all variables *multiple nominal*. This is what we have described earlier as simple homogeneity analysis or multiple correspondence analysis. In Figure 3.4 we give a two-dimensional principal component analysis representation of our small example, using the geometry or homogeneity analysis.

Figure 3.4 results from analysing Table 3.2. It is clear, of course, that the analysis of Tables 3.1 and 3.3 would give different results in general. Table 3.3 is quite interesting in this respect. For Table 3.3 the indicator matrices G_j are permutation matrices. If we substitute them in (2.1) it is obvious that loss can always be made equal to zero by letting X be an arbitrary $n \times p$ matrix, and by setting $Y_j = G_j'X$. Then $G_j Y_j = G_j G_j' X = X$. In the same way single nominal variables can always be fitted perfectly. Choose X and a_j arbitrarily, and set $z_j = G_j' X a_j$. Then $q_j = X a_j$, and loss is minimized by (4.2). In other words:

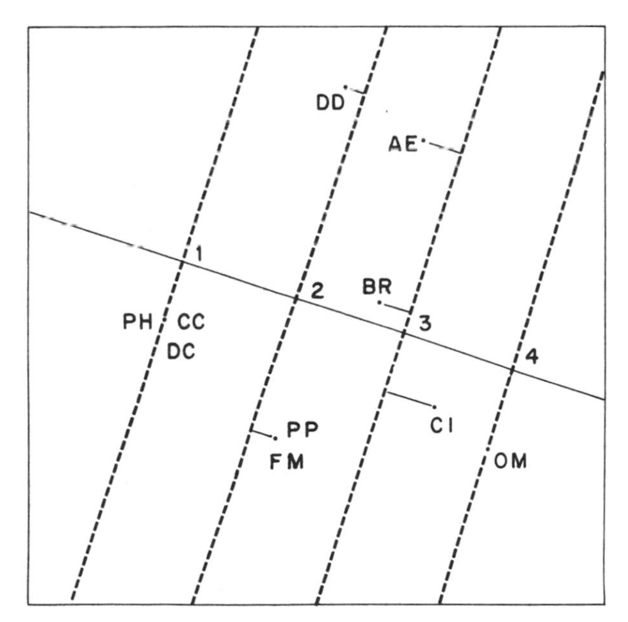

Figure 3.4. Single numerical loss, variable 1, optimal solution

non-trivial analysis of rankings is possible only if we make all variables either single ordinal or single numerical. It is also interesting to compare the single quantifications in $q_j = G_j z_j$ with the original scores in Table 3.1. Clearly plotting the elements of q_j versus the original scores will give a step-function. We have discreticized our variables, and as a consequence every object in the same discretization interval gets the same quantification in the q-vector of the variable. The more intervals, the less crude the transformation given by the step-function will be, but no matter how fine we choose the discretization, the transformation will always be a step function. And step functions do have several drawbacks as we can verify in Chapter 2. This is one of the main reasons why we say that homogeneity analysis as currently implemented by Gifi (1981b, 1982) has a discrete bias. Step-functions are perfectly natural for variables which have a small number of possible values to start with, or for purely nominal variables for which we have no prior numerical information. For 'continuous' numerical variables, such as the three variables in our example, transformation by step-functions ignores the prior information that our variable was originally continuous, and can also assume all intermediate values between the end-points. Thus we now know how to incorporate numerical and ordinal information, but we do not know yet how to incorporate 'smoothness' into homogeneity analysis. This problem will be discussed below, but first we have to fill a number of gaps that have been left open in the combination of various options we have discussed up to now.

6. GAPS IN GIFI

In the previous sections we have discussed single numerical, single ordinal, single nominal, and multiple nominal variables. We did not discuss multiple ordinal and multiple numerical. If only for aesthetic reasons it is interesting to investigate if these remaining types of variables can also be given a simple meaning. Moreover we have distinguished single and multiple variables. For single variables we required that rank (Y_j) was less than or equal to one, for multiple variables there were no rank restrictions, which means that we 'required' that rank (Y_j) was less than or equal to min $(p, k_j - 1)$. It is $k_j - 1$ and not k_j in this upper bound, because of the fact that the rows of Y_j have a weighted mean of zero. Now if $p = 1$ there is no difference between multiple and single. If $p = 2$ then for variables with more than two categories single requires that rank (Y_j) is less than or equal to one and multiple that rank (Y_j) is less than or equal to two. There is no gap between the two options. But for $p = 3$, and k_j larger than three, single requires rank one and multiple requires rank three as the upper bound. Thus there is a gap. We can insert another option, which requires rank (Y_j) to be less than or equal to two. This general rank restriction, which can be between single and multiple, was already discussed in de Leeuw (1976), but it was not incorporated in the subsequent developments of the Gifi system.

The loss function, with general rank constraints, can be written as

$$\sigma(X; Y_1, \ldots, Y_m) = \sum_j \operatorname{tr}(X - G_jZ_jA_j')'(X - G_jZ_jA_j'). \tag{6.1}$$

Here Z_j is $k_j \times r_j$, and A_j is $p \times r_j$. The r_j are the required ranks for variable j. Geometrically the constraint means, of course, that the category quantifications must be in a r_j-dimensional hyperplane through the origin. If $Z_j'D_jZ_j = nI$, then loss for variable j satisfies

$$\sigma_j(X, Y_j) = n(p - r_j) + \operatorname{tr}(XA_j - G_jZ_j)'(XA_j - G_jZ_j). \tag{6.2}$$

If $A = (A_1 \mid \ldots \mid A_m)$ and $Q = (Q_1 \mid \ldots \mid Q_m) = (G_1Z_1 \mid \ldots \mid G_mZ_m)$, then

$$\sigma(X; Y_1, \ldots, Y_m) = nm(p - r) + \operatorname{tr}(XA - Q)'(XA - Q). \tag{6.3}$$

This looks very similar to (4.2), but remember that in (6.3) each Q_j consists of r_j orthogonal quantifications of the same variable, i.e. of r_j *copies* (compare de Leeuw, 1984a; Tijssen, 1984; de Leeuw and Tijssen, 1984). Again, geometrically, we have minimum loss if the category points are in an r_j-plane, and all object points are on lines perpendicular to the plane, which cross the plane in the k_j-category points.

General rank restrictions now make it possible to define r_j-nominal, in which there are no further restrictions on Z_j. There is also r_j-numerical, in which the r_j columns of Z_j are known orthogonal k_j-vectors. And, finally, there is r_j-ordinal, in which all columns of Z_j must be in the appropriate order. For r_j-nominal and r_j-numerical we can require, without loss of generality, that $Z_j'D_jZ_j = nI$. For r_j-ordinal such a constraint cannot be imposed, and we have to refrain from normalizing Z_j and/or A_j. It is clear, of course, that general rank constraints, coupled with *measurement restrictions*, generalize our previous notions of single and numerical, and fill the gaps in the system. In fact it opens completely new possibilities: we can require that the first 'copy' in Z_j is ordinal, while the remaining copies are nominal, and so on. Again we do not know how practical these new options are. We have discussed them because they fit naturally into the gaps, and also because they can be incorporated without much ado into the homogeneity analysis algorithms that are already there.

7. PSEUDO-INDICATORS

In Chapter 2 it is illustrated that a more satisfactory analysis of continuous variables becomes possible if we generalize the notion of an indicator matrix. Suppose we continue to use the same notion of loss, with the same types of restrictions on the category transformations, but we do not suppose that the G_j are indicator matrices. They must still be known $n \times k_j$ matrices, but they need not be binary any more. In a sense we have already gone a step in this direction. If a variable is r_j-numerical, then $Y_j = G_j(Z_jA_j') = (G_jZ_j)A_j'$. Suppose, for instance, that the Z_j are polynomials, orthogonal with respect to the marginals. Then G_jZ_j are orthogonal polynomials in n-space, and we can interpret our analysis as an unrestricted analysis using an $n \times r_j$ basis of orthogonal polynomials instead of the indicator matrix G_j. Although this is clearly a valid interpretation, it is not exactly what we have in mind.

In this section we concentrate on so-called *fuzzy codings*, collected in *pseudo-indicator* matrices. Indicator matrices are characterized as pseudo-indicators with bandwidth unity, cf. Chapter 2. Piecewise linear B-splines define pseudo-indicators with bandwidth two, and so on. In this chapter we do not care about the origin of the pseudo-indicators, for this we refer to Chapter 2. We simply assume that data are coded in this way, and we look for the geometrical interpretations of such a coding. In Table 3.4 we have a fuzzy coding of our small

Table 3.4. Piecewise linear coding car data

	Price			*Gas*			*Weight*			
Chevette	0.88	0.12	0.00	0.62	0.38	0.00	0.06	0.94	0.00	0.00
Dodge Colt	0.86	0.14	0.00	0.98	0.02	0.00	0.24	0.76	0.00	0.00
Plymouth Horizon	0.74	0.26	0.00	0.90	0.10	0.00	0.02	0.98	0.00	0.00
Fort Mustang	0.48	0.52	0.00	0.66	0.34	0.00	0.00	0.60	0.40	0.00
Pontiac Phoenix	0.28	0.72	0.00	0.62	0.38	0.00	0.00	0.58	0.42	0.00
Dodge Diplomat	0.12	0.88	0.00	0.00	0.96	0.04	0.00	0.00	0.90	0.10
Chevrolet Impala	0.00	0.98	0.02	0.50	0.50	0.00	0.00	0.00	0.62	0.38
Buick Regal	0.00	0.90	0.10	0.44	0.56	0.00	0.00	0.00	1.00	0.00
AMC Eagle	0.00	0.86	0.14	0.00	0.66	0.34	0.00	0.00	0.86	0.14
Oldsmobile 96	0.00	0.34	0.66	0.26	0.74	0.00	0.00	0.00	0.34	0.66

example, which is actually the result of piecewise linear coding. The idea behind our generalization of homogeneity analysis now is, that we can combine all our previous options and restrictions with this new coding as well.

In particular we can impose rank-constraints, and impose ordinal or numerical restrictions.

Because $p=2$ in our example it suffices to distinguish single and multiple. Consider multiple nominal. The loss component for variable j vanishes if $X=G_jY_j$. In the coding used in Table 3.4 each X corresponds with two categories, because the bandwidth in our example is two. The two category quantifications are the endpoints of a line segment, all line segments for a particular variable are connected. The object scores must be on the line segment corresponding to the categories they are in. And not only must they be on the segment, they must also be in a precise location on the segment, where the location is dictated by the masses of the endpoints in the coding. This is indicated in Figure 3.5, which is not an optimal solution of any kind, but it is used to illustrate the loss of variable 1 in the coding of Table 3.4. The points on the two line segments indicate where the

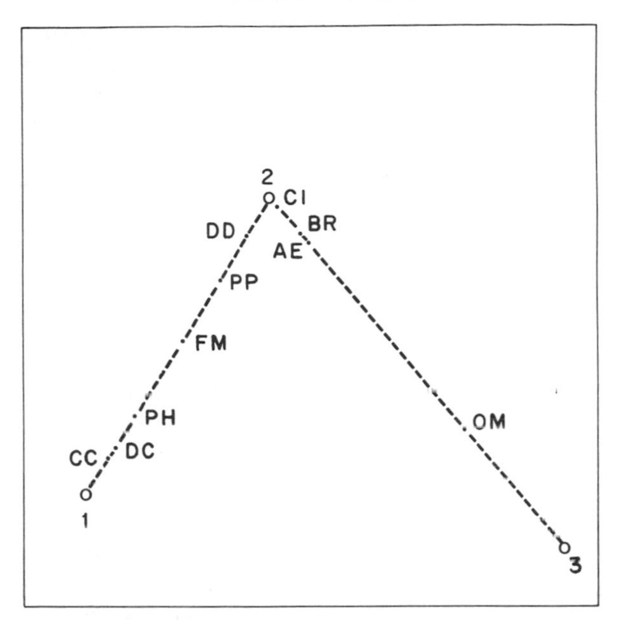

Figure 3.5. Loss for multiple piecewise linear, variable 1, arbitrary solution

cars must be given the coding, and given the location of the endpoints of the segments. In the single case the endpoint must be on the same straight line, and the object points must project on the places fixed by the coding. Thus there are parallel lines perpendicular to the line connecting the category points, which intersect this line at the appropriate places. In the ordinal case the endpoints must be ordered along the line, such that both within-category and between-category quantification is ordered (which makes this a somewhat peculiar option, perhaps).

The geometry of fuzzy or smooth homogeneity analysis with several applications is extensively treated in van Rijckevorsel (1987). If we study the transformation which considers $q_j = G_j z_j$ as a function of the original data values, then transformations from pseudo-indicators as discussed in Chapter 2, will indeed be more smooth than those from indicators. The precise nature of the smoothness depends on the nature of the pseudo-indicators, for instance on the bandwidth. In our example the transformations are continuous and piecewise linear. If we use piecewise quadratic splines, joined in such a way that they are differentiable at the endpoints, then we get more smoothness (and a bandwidth of three). The geometry becomes more complicated, because object scores must be at the appropriate places in the triangle spanned by three endpoints. Successive triangles are interlocked, because they have one side in common. And so on, for larger bandwidths, and/or in higher dimensions. The relationship of the object scores with the multiple quantifications (Y-configuration) and its interpretation are to be reconsidered. Several questions do arise: What happens to the *principe barycentrique* in fuzzy homogeneity analysis? Are the original (=not coded) data reproducible from the final configuration? What is the geometrical significance of the goodness of fit parameters?

7.1. The representation of basis functions

First order B-splines (crisp coding) collapse all values within an interval into a point. Second order B-splines force all values within one interval to be on a line. Third order B-splines transform all values to be on the face of a triangle.

The regular polygon is often used as a triangle of reference for the position of data points in the basis, cf. Le Foll (1979), Gallego (1980), Greenacre (1984) and van Rijckevorsel (1987). In this way we can represent at most three (orthogonal) dimensions, i.e. basis functions, in a plane. This implicitly uses the property that all fuzzy codes of one data point add up to one. The use of triangular coordinates is limited to the representation of three basis functions at the time.

This tool enables us to show the differences between various low dimensional forms of fuzzy coding in terms of triangular coordinates. The triangle of reference with the vertices (1, 0, 0), (0, 1, 0) and (0, 0, 1) is also known as the triangular or barycentric coordinate system. The number of coordinates $\neq 0$ is maximally three per data point. In Chapter 2 this number is also called the bandwidth of the set of basis functions.

In the cases (a) and (b) in Table 3.5, one triangle suffices to represent the whole transformation function because there exist only three basis functions, cf. Figure 3.6: A, B_1 and B_2.

The crisp codes in Figure 3.6 coincide with the vertices. The codes in the fuzzy areas around the knots are the points on the sides of the triangle. The transition from first order fuzzy coding into second order codes is clearly because in the latter case all codes are between the vertices. This automatically leads to second order B-splines, where all points are on the sides between the vertices and only coinciding with a vertex, if the data-point coincides with a knot.

Table 3.5. Dimensionality, bandwidth and order of low order fuzzy coding with two interior knots

Type of coding	*Order*	*Dimension*	*Bandwidth*
(a) Crisp coding	1	3	1
(b) First order fuzzy coding	1	3	2
(c) Second order B-splines	2	4	2
(d) Third order B-splines	3	5	3

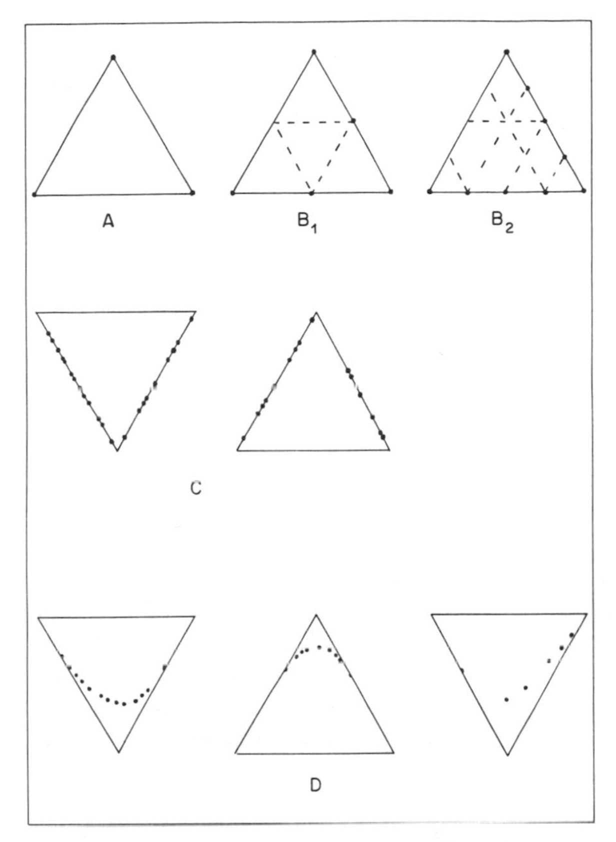

Figure 3.6. The Public spending in the Netherlands between 1951 and 1981 coded by crisp coding (A), demi-discrete (B), trapezoidal (B2), piecewise linear (C) and quadratic coding (D), all represented by triangular coordinates after van Rijckevorsel (1987), see also Chapter 2

The first degree fuzzy coding represents three intervals that have two coordinates $\neq 0$ on four basis-functions, cf. Figure 3.6 (C). One triangle is not sufficient because of the dimensionality of the basis. One way of solving this is by using an additional triangular coordinate system that has one dimension, i.e. one side between two vertices, in common with the first triangle, in order to maintain the simplicity and parsimony of this approach. The first triangle covers the first two (out of three) intervals and the second triangle the last two (out of three) intervals. There exists an overlap of one interval, and they have two vertices in common; i.e. the triangles are interlocked.

The second degree coding of Figure 3.6 (D), is a code with three different coordinates $\neq 0$, that add up to one, which is represented by a triangle of reference in Figure 3.7.

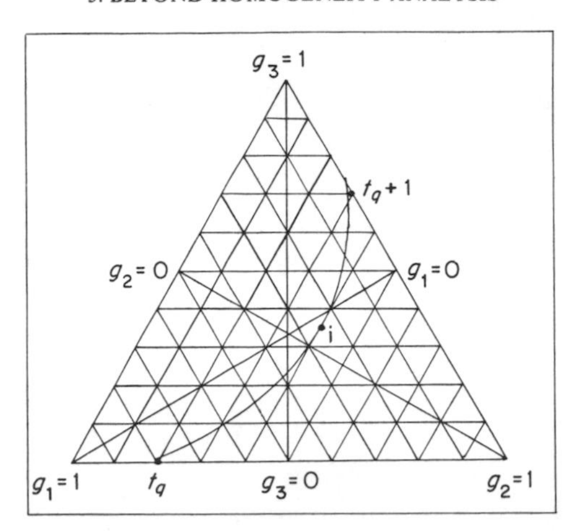

Figure 3.7. The fuzzy coding of data point i by second degree B-splines: $\{g_1=0.25, g_2=0.40, g_3=0.35\}$, represented by the barycentric coordinate system

It is clear that, if one of the three coordinates equals zero, the corresponding point is on the side of the triangle. The coordinates of the knots are t_q: $\{g_1=0.8, g_2=0.2, g_3=0.0\}$ and t_{q+1}: $\{g_1=0.0, g_2=0.7, g_3=0.3\}$. Note that in every knot one of the codes is equal to zero. *Ergo* the knots are on the sides of the triangle and the points within the interval are on the face of the triangle. All data values between t_q and t_{q+1} are on the quadratic curve between t_q and t_{q+1}. Each basis vector is a quadratic function and hence the triangular representation is a quadratic as well.

In this way the generalization to second degree codes, cf. Figure 3.6 (D), is easy to understand. The restricting parameter has evolved from a point, via a line segment, to the face of a triangle. A second degree B-spline is a quadratic function on the face of the triangle of reference smoothly joining at two sides.

7.2. The build-up of a transformation function

In crisp coding data points are represented as grouped points, by first degree B-splines as individual points on line segments and by second degree B-splines as points on curves. We know from Chapter 2 that the global transformation function is equal to a piecewise function.

Say, we use a triangle of reference to represent the coding of a variable. Then we can observe the same phenomena, i.e. point, line segment and curve in the barycentric representation. If we inspect the functional coefficients in the space of the object scores, which is the usual way of inspecting the parameters in homogeneity analysis, we observe the same phenomena: point, line segment and curve. This is to be expected; the sets of functional coefficients span the subspaces that geometrically restrict the object scores. In case of perfect fit the object scores collapse, not into the functional coefficients, but into the regular polytopes spanned by these coefficients. They span points, line segments and curves on the face of a triangle. This means that a perfect fit in fuzzy homogeneity analysis is trivial in a different way from a perfect fit in crisp homogeneity analysis. The latter demands that all data points collapse perfectly into a few categories. This seems less realistic than demanding that all data points within an interval should form a linear or quadratic function, which permits the expression of a considerably larger amount of variation in a controlled way (see Figure 3.8).

Note that the transformation $G_j Y_j$ coincides with the functional coefficients in crisp coding, it connects the coefficients in first degree B-splines by straight lines and in second degree B-splines it forms a quadratic curve that contains only those functional coefficients that correspond to the exterior knots.

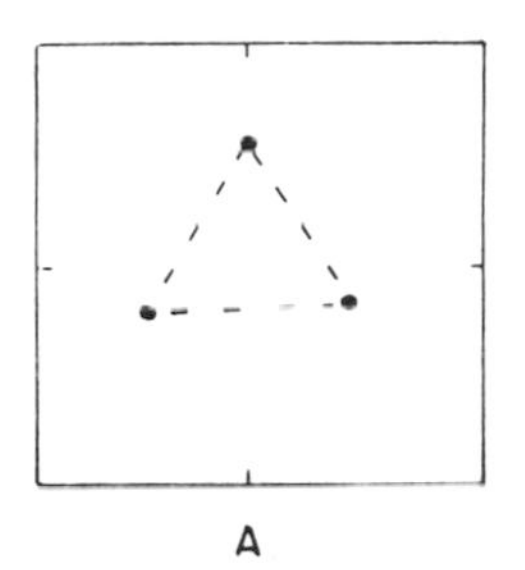

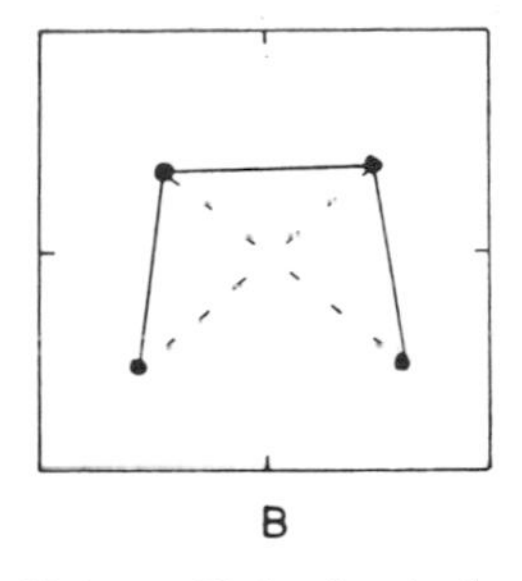

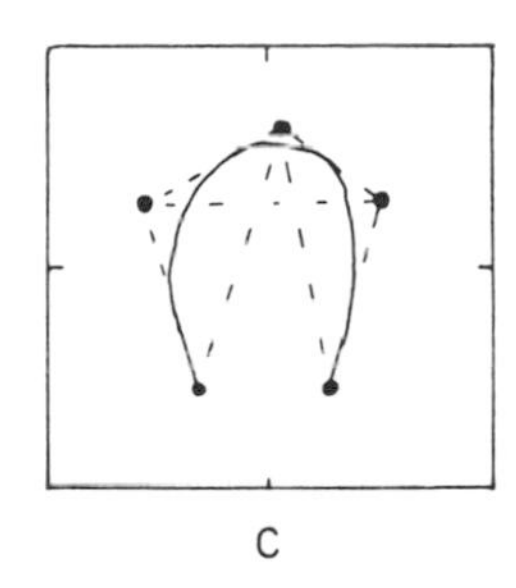

Figure 3.8. The functional coefficients Y_j (=dots), the weighted triangles of reference (=dotted lines) and the global multiple transformation functions $G_j Y_j$ (=solid lines) in the X-space. Represented for crisp coding (A), first degree B-splines (C) and second degree B-splines (D)

7.3. Goodness of fit

The goodness of fit of point i on variable j is defined as is the custom in homogeneity analysis: the squared euclidean distance between the observation score x_i and the corresponding value on the global multiple transformation function $g_{ij} Y_j$. See the dotted lines in Figure 3.9.

The subspaces, spanned by the functional coefficients, can be interpreted geometrically as restrictions for the corresponding object scores. Using crisp codes, the objects scores should be as close as possible to the point Y_{jk}; using hat codes (=second order B-splines), the object scores should be as close as possible to the corresponding points on the line segments between the points $Y_{j,k}$ and $Y_{j,k+1}$; as for the bell codes (=third order B-splines) the object scores should be as close as possible to the curve within the face of the triangle, spanned by the functional coefficients $Y_{j,k}$, $Y_{j,\ k+1}$ and $Y_{j,k+2}$. We can now develop a schematic geometrical account of what happens to a variable in fuzzy homogeneity analysis.

The data are coded respectively by all five nearly-orthogonal fuzzy codes discussed in Chapter 2. The symbols used in the Figures 3.10 to 3.13 are A (crisp),

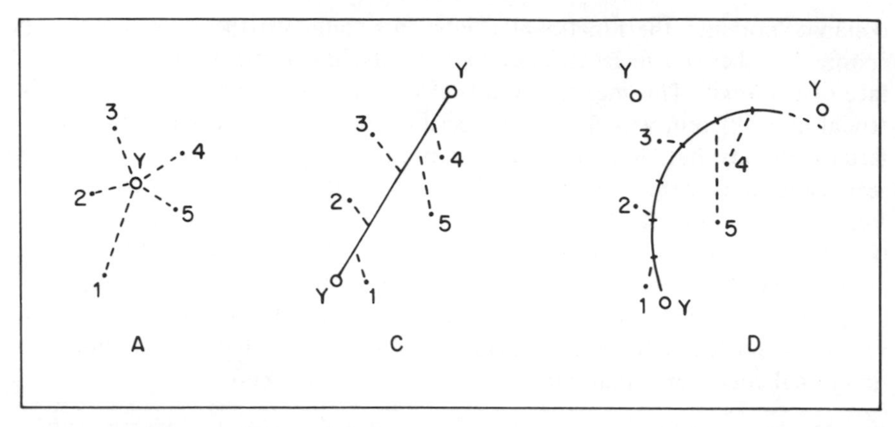

Figure 3.9. The euclidean distances between x-points and the global multiple transformation function in one interval for three different ways of coding

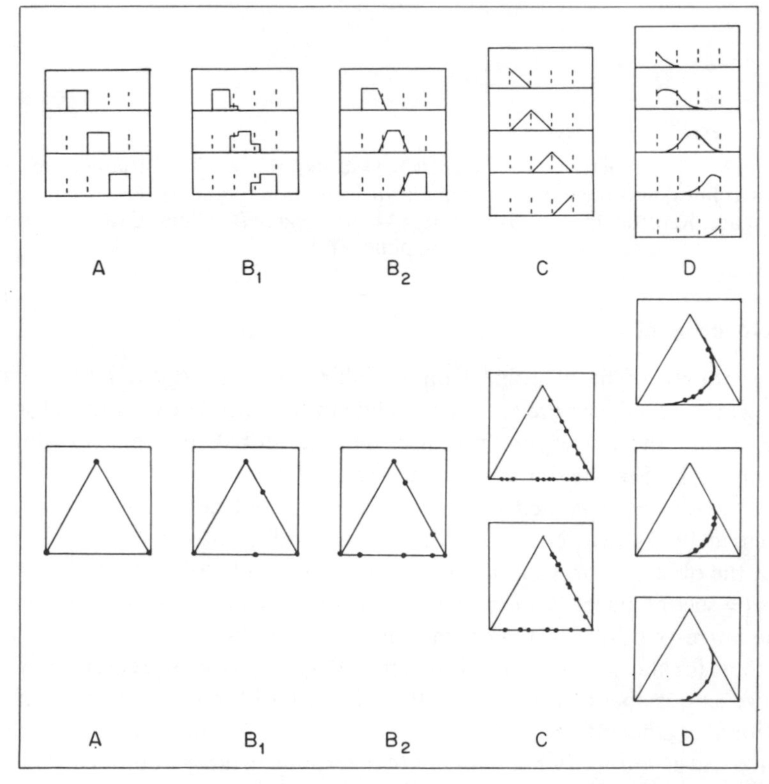

Figure 3.10. Five different types of fuzzy coding functions with the representations by triangular coordinates; see the text for the interpretation of the labels

B1 (semi-discrete), B2 (trapezoidal), C (first degree B-spline) and D (second degree B-spline). The obtained basis vectors are represented by triangles of reference: maximally three basis vectors per triangle. A triangle of reference can represent three intervals in case of zero degree codes, two intervals in case of first degree codes and one interval in case of second degree codes (and no intervals for higher degree codes).

The basis vectors, and thus the vertices of the triangles of reference, are weighted with respect to the p-dimensional X configuration for maximal homogeneity by the least squares estimates $Y{:}G_jY_j$, while $G_{jk}Y_j$ is the quantification of data points in the kth interval, see Figure 3.11.

The weighted basis functions expressed by the sides of triangles, form together the global multiple transformation function in the p-dimensional space, see Figure 3.12. Nothing new is introduced here. The procedure is extensively discussed in Chapter 2 and in this chapter. Each side of the triangle is geometrically speaking, separately stretched respectively shrunk, by the least squares estimation.

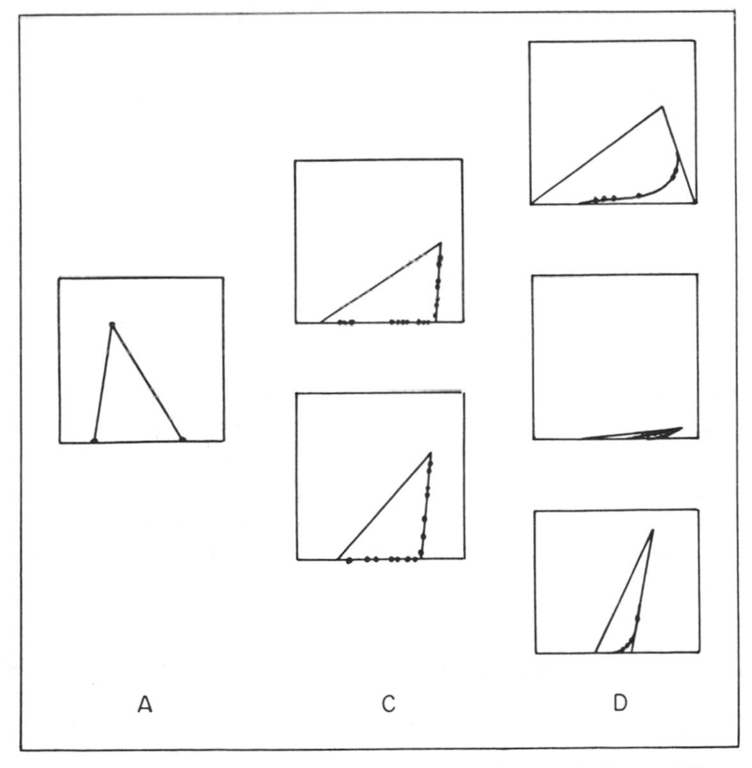

Figure 3.11. The weighted triangular representations of three different types of fuzzy coding: see the text for the interpretation of the labels

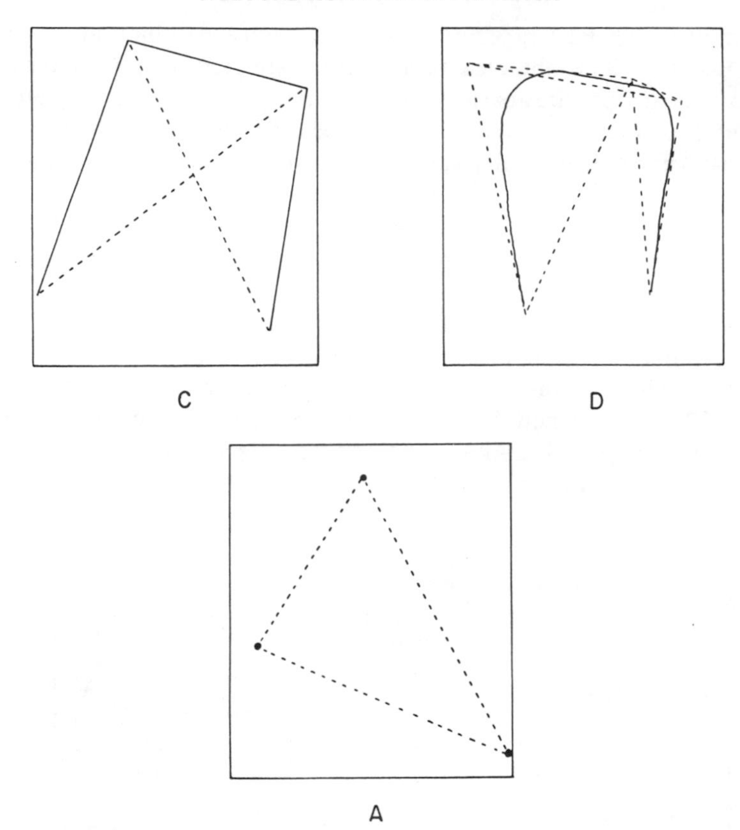

Figure 3.12. Weighted triangular representations (dotted) and resulting multiple transformation functions (solid) in the first two dimensions of fuzzy homogeneity analysis, based on three different types of fuzzy coding

The weighted triangles have a distinct relationship with the X-configuration: It follows from this picture that bandwidth three or more does not combine naturally with single quantification, because single quantification makes the triangles degenerate to straight lines. This is no problem analytically, but it makes the geometry of loss far less interesting. In general we think that for practical purposes a bandwidth larger than two is probably not very interesting, unless data are very well behaved indeed.

8. RELATED WORK ON FUZZY HOMOGENEITY ANALYSIS

The combination of homogeneity analysis and fuzzy coding is fairly recent. The development of fuzzy set theory, comprehensively reviewed by Bezdek (1987), took place independently from the development of homogeneity analysis. See also van Rijckevorsel (1987). Fuzzy coding itself is introduced by Zadeh (1965)

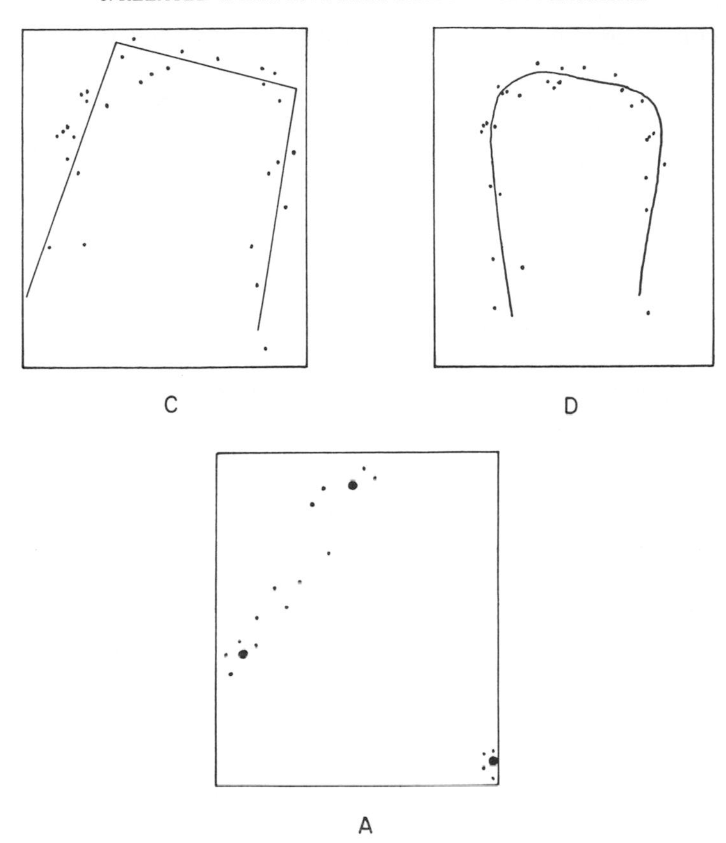

Figure 3.13. Multiple transformation functions and X-points in the plane of the first two dimensions of fuzzy homogeneity analysis based on three different types of fuzzy coding

and Ruspini (1969) and the first proposals for fuzzy coding in homogeneity analysis are by Bordet (1973), Guitonneau and Roux (1977) and Ghermani, Roux and Roux (1977). These early papers mainly adapt continuous data to the discrete mould of homogeneity analysis and hence try to smooth the inaccuracy of crisp coding around the knots. The application within homogeneity analysis seems to be rather accidental; theoretically speaking it could have been any other crisp technique. In the same vein Martin (1980) derives a probabilistic framework to this end. See also Chapter 5. The first attempts to incorporate fuzzy coding and homogeneity analysis are by Le Foll (1979) and Gallego (1980). They both concentrate on piecewise linear coding. Le Foll generalizes to a larger class of coding strategies, which he defines as *codages complets* to be used in a variety of techniques. The larger part of his work, however, is devoted to an application of piecewise linear coding within homogeneity analysis on ecological data, referring to the surface water pollution in the larger surroundings of Paris (France).

Gallego (1980) concentrates on an application of piecewise semi exponential coding within several techniques in order to smooth seasonal macroeconomic data. In this context he discusses linear PCA (on recoded data) and cluster analysis but predominantly homogeneity analysis.

Some French work on fuzzy coding is marked by the desire to analyse real valued and categorical data simultaneously within one analysis. Fuzzy coding is thus a means to incorporate real valued data by coding them into a format that conforms with homogeneity analysis. And, consequently, because fuzzy coding in itself increases inertia, much attention is given on methods how this should be corrected for, prior to further analysis. cf. Guitonneau and Roux (1977) Benzécri (1980), Gallego (1980) and Greenacre (1984, p. 162).

Another group of French authors prefer to work with a probabilistic interpretation of fuzzy coding in non-linear analysis, cf. Chapter 5. Martin (1980) is most outspoken and other work often uses his definition of probability coding, that a fuzzy code is mainly a transition probability between an observed and an unobserved variable, cf. Besse and Vidal (1982), Gautier and Saporta (1982) and Mallet (1982). The idea is that the unobserved random variable is reconstructed by coding the observed variable. Besse and Vidal (1982) extend this idea to both variables, observed and unobserved, being categorical and restrict the coding to the bivariate coding of pairs of variables (not to be confused with the bivariate coding of pairs of intervals as mentioned in Chapter 2). Ramsay (1982) and Besse and Ramsay (1986) discuss the (smooth) PCA of data which are functions, cf. Chapter 4. This work should not be confused with the work of Winsberg and Ramsay (1983) who consider the isotone polynomial spline transformations of separate variables as a kind of (probabilistic) optimal scaling. See also the ACE methodology applied in this context by Koyak (1985), and the work of Winsberg and Kruskal (1986). Chapter 6 deals separately with the latter.

Apart from these developments there exists another tendency to relate fuzzy coding within homogeneity analysis to the theoretical non-linear principal component analysis as defined by Dauxois and Pousse (1976). Fuzzy coding is then regarded as a way to further the convergence of homogeneity analysis to a theoretically completely non-linear generalized canonical analysis, where non-linear variables are non-linearly related. This is linked to the two types of convergence discussed in Chapter 2. Nearly all the French research in this field published after 1976 refers to this form of non-linear generalized canonical analysis. See also Lafaye de Micheaux (1978) and Mallet (1982). The latter conjectures that the empirical analysis of fuzzy-coded variables is a good approximation of the theoretical non-linear analysis.

9. PROCESS

In the developments so far the data were coded as (pseudo)-indicators, and these pseudo-indicators were fixed during the computations. Now let us look at single ordinal piecewise linear again. We have already seen that the order of the category points on the line is fixed in this case, although their precise location is free. Given the location of the category points, however, the location of the preferred projection of the objects points on the line is fixed by the coding. This is what we mean by fixedness of within-category order. This fixedness is contrary to what is called the *primary approach to ties* in multidimensional scaling literature, and also the *continuous ordinal* option (compare de Leeuw, Young and Takane, 1976, Young, de Leeuw and Takane, 1980, Young, 1981). In this option, which is incorporated in various non-metric principal component programs, we fix the order between categories but not within categories. Or, geometrically, given the line and the location of the category points on the line, the object-point can project *anywhere* between the end-points of its category. Loss only occurs if they project outside their assigned interval.

Given our previous discussion it is easy to see how the idea of continuous ordinal data can be incorporated easily into our form of homogeneity analysis. The elements of the pseudo-indicators are not considered fixed any more, only the location of the non-zero elements is fixed. Thus we know which elements must be non-zero, we also know that they must be non-negative and they must add up to one for each row, but their precise values are additional parameters over which the loss function is minimized. In the single ordinal piecewise linear case this gives exactly continuous ordinal data as treated in PRINCIPALS, for instance (Takane, Young and de Leeuw, 1978). But because we have fitted the possibility of varying the elements of the G_j into our general homogeneity analysis framework, we can combine this option with all other previous options that we already had. It can be combined with multiple quantification, and with single numerical quantification. In this last case it gives the continuous numerical scaling earlier discussed by de Leeuw and Walter (1977).

There is very little need to elaborate on the geometry of the continuous versions. It is basically the same as the discrete geometry, only points are not fixed in intervals, but they can be anywhere in the interval. It becomes perhaps a bit more interesting to use larger bandwidths with single options, because the bandwidth now controls the amount of overlap of the intervals corresponding with the categories. If bandwidth is two, there is no overlap. If bandwidth is three, successive categories have one common subinterval, and so on. Multiple options with bandwidth three, in two dimensions, are interpreted in terms of triangles (or convex hulls). Objects in category 1 must be in the convex hull of category points 1, 2 and 3, objects in category 2 in the convex hull of 2, 3 and 3, and so on. Successive triangles have one side in common, if they degenerate to line segments this becomes the overlapping subinterval. It is not at all clear (yet) if these conceptually very nice options are useful in practice. A theorem in Gifi (1981a, 1988) is useful to illustrate their limitation. It refers to the continuous ordinal option, with all variables single. The results show that with this option *degenerate* solutions, which locate one object very far away from the others, which are collapsed into a single point, will be quite common. In fact Gifi shows that in the situation in which objects are a random sample the minimum of loss is almost surely equal to zero if the sample size tends to infinity. Van Rijckevorsel (1987) illustrates with real data that the conceptual nicety can be misleading. We do not know yet how devastating the results are in practice, but it certainly indicates that we have to be careful.

Computationally our new options do not introduce any trouble at all. We must introduce a new subproblem into the alternating least squares cycles of homogeneity analysis in which the G_j are adjusted. This is done for each row of each G_j separately, defining a very small special quadratic programming problem. Of course we have to exert a little self-control in combining our options. We have the possibility, in principle, to take a different bandwidth for each object, or a different rank for each Y_j. In fact, looming large in the distance, is the possibility of further generalizations. We can fix the bandwidth of each variable, for instance, and determine the optimum location of the non-zero elements. This is probably very unwise, because the program output will become almost independent of the data.

It is perhaps convenient to relate existing programs to our general form homogeneity analysis, in which we choose (a) quantification rank, (b) measurement level, (c) bandwidth, (d) process for each variable separately. HOMALS (Gifi, 1981b) has quantification rank equal to dimensionality, measurement level nominal, bandwidth unity and process discrete. Of course if bandwidth is unity there is no distinction between discrete and continuous process. Ordinary principal component analysis has quantification rank unity, measurement level numerical, bandwidth unity and process discrete. PRINCALS (Gifi, 1982) has quantification rank either one or dimensionality, and measurement level numerical, ordinal or nominal (but ordinal/numerical cannot occur together with multiple), bandwidth is unity and process is discrete. SPLINALS (van Rijckevorsel, 1982, 1987; Coolen, van Rijckevorsel and de Leeuw, 1982) has quantification rank either one or dimensionality, measurement level nominal, bandwidth either one or two and process discrete. Winsberg and Ramsay (1983) have, with some minor qualifications, measurement level ordinal, quantification rank unity, arbitrary bandwidth, and process discrete. PRINCIPALS (Takane, Young and de Leeuw, 1978) has quantification rank-one, measurement level nominal, ordinal or numerical, bandwidth either one or two, process continuous or discrete. But if the process is continuous the measurement level must be ordinal, and if the process is discrete the bandwidth must be one. It is clear that our new homogeneity analysis program, which only exists in preliminary APL-versions yet, encompasses all these possibilities and has all previous programs as special cases. Of course it will be more expensive in terms of time and storage, and more liable to produce degeneracy.

10. WORDS OF CAUTION

Homogeneity analysis is a dangerous technique. We use very little information from the data, and we do not impose restrictions of a strong type on the representation. This type of program traditionally appeals greatly to many social scientists, who are very unsure about the value of their prior knowledge. They prefer to delegate the decisions to the computer, and they expect programs to generate knowledge. This strategy leads, all too often, to chance capitalization, triviality and degeneracy. Hypotheses are never rejected, and investigators are constantly making errors of the second kind. As a consequence results can, of course, never be replicated. Generalized homogeneity analysis, as we have developed it here, is a very powerful tool which can contribute greatly to a further inflation of social science results. By choosing the least restrictive options we can make the results almost completely independent of the data.

On the other hand it is well known that if we pay too much attention to errors of the second kind, then social scientists can say abolutely nothing. This is also considered to be an undesirable state of affairs. It can be circumvented by concentrating on minute aspects of well-defined small problems, as in laboratory situations, or it can be circumvented by introducing vast quantities of prior knowledge, as in sociology. Of course in most cases the prior knowledge is nothing but prejudice, and it so dominates the investigation that the results become equally independent of the data.

This defines the dilemma of applied empirical social science. According to the canons of scientific respectability we can say almost nothing, and the things we can say are likely to be trivial. There are two ways out of this situation. Either we impose so much prior knowledge on our problem that the data only marginally make a difference. This is the rationalistic solution, popular in sociology. Or we impose so little prior knowledge that the data, including all outliers, stragglers, idiosyncrasies, coding errors, missing data, completely determine the solution. In this case the technique is supposed to generate theory. This is the empiristic and technological approach, popular in applied psychology. Both approaches have, up to now, not produced much of interest.

Homogeneity analysis is firmly in the empiristic and technological tradition. Thus it is clear what dangers we have to guard against especially. If we have reliable prior knowledge, we must incorporate it. It is absolutely necessary to investigate the *stability* of the results (Gifi, 1981a; de Leeuw, 1984b). Observe, however, that stability is not sufficient. A program that responds to any data matrix by drawing the unit circule is very stable indeed. We also need to *gauge* the technique, by comparing analysis with different options on data whose most important properties are known. For some forms of homogeneity analysis this has already been done quite extensively (Gifi, 1981a; de Leeuw, 1984c), but apart from results by van Rijckevorsel (1987) very little is known in this respect about the more general options discussed here. One strategy, that seems promising, is to analyse the same data with various options, from numerical to ordinal, from bandwidth one to bandwidth two, from discrete to continuous, and so on. In fact, this defines another form of stability analysis, which seems indispensable in situations with little prior knowledge.

Component and Correspondence Analysis
Edited by J.L.A. van Rijckevorsel and J. de Leeuw

Chapter 4

Spline Functions and Optimal Metric in Linear Principal Component Analysis

Philippe Besse
Paul Sabatier University, Toulouse, France

1. INTRODUCTION

Spline functions are now widely used in various fields of statistics with different objectives. Wegman and Wright (1983) review these developments in the frame of non-parametric estimation. We focus here on applications in the field of exploratory data analysis. First we briefly recall the differences between non-linear and linear approaches of data analysis, then we propose a strategy specially adapted to longitudinal data. It is based on the use of metrics induced by Sobolev spaces on spline interpolation subspaces. Two rationals led to this choice of functional framework:

—Some good properties of convergence (Besse, 1980),
—In practice, it is very easy to deal with spline functions.

1.1. Non-linear analysis

Let us consider a set of data consisting of values of n objects with respect of p variables displayed in a $n \times p$ matrix. Three approaches lead to comparable results:

(a) Winsberg and Ramsay (1983) are looking for monotone transformations of the data matrix columns to achieve a dimension reduction in PCA (see Chapter 6). These transformations are approximated by I-spline functions.
(b) de Leeuw *et al.* (1981) follow the same purpose but use B-spline functions to define transformations, which are not necessarily monotone, and optimize an

alternative least square algorithm rather than a likelihood function (see Chapters 1, 2 and 3).

(c) The approach of Lafaye de Micheaux (1978) or Martin (1980) is developed within the framework of semi-linear PCA (Dauxois and Pousse, 1976) and can be seen as a particular case of (b) above (see also Chapter 5).

Let X be a random variable mapping a probability space (Ω, α, μ) into $(\mathbb{R}^m, \mathscr{B}_{\mathbb{R}^m})$. We can briefly sum up the work above by writing that they lead to seek for normalized f_i functions in some finite-dimension subspaces L_i of $L^2(\mathbb{R}, \mathscr{B}_{\mathbb{R}}, v_{x_i})$ that maximize the sum of the p largest eigenvalues ($p=1$ in case (c)) of the linear PCA of $\{f_i \circ X_i; i=1, \ldots, m\}$. A monotonicity condition on the f_i restricts the L_i space to a cone and the main problem is to choose the bases defining the L_i spaces. Classical analyses use characteristic functions of intervals ('crisp coding'). However, the above-mentioned authors use some piecewise polynomial functions (spline). This requires beforehand to choose the regularity of the transformations (order of splines), the number and the situations of the knots for each variable.

1.2. Linear analysis

Spline functions are used in the non-linear analysis framework to approximate non-linear (monotone or not) transformations of each variable. In other work (Besse, 1980; Saporta, 1985; Besse and Ramsay, 1986) spline functions are used in the different context of linear PCA. These papers, and this chapter, describe a PCA technique for data consisting of n functions each observed at p arguments, for instance n time series. In this case the problem is to take into account the time or, rather, the order structure of the variables. This requires to consider each object not only as a set of values but as the approximation of a real function. Then, the aim of PCA is to display the respective *shapes* of the curves in a scatter diagram. It is adaptatively achieved by taking into account first or higher derivative characteristics to measure dissimilarities between curves. In other words, following some investigations, the relevant information can be found by comparing growths or convexities and not only values.

This technique can be performed by an approximate choice of metric or, equivalently, by a linear transformation of the data matrix rows prior to a conventional PCA. It requires two decisions:

(a) How regular are the curves or trajectories? It means: how many derivatives are the functions being sampled presumed to have?

(b) How will these derivatives as well as the functions themselves be used in order to assess the similarity of two functions?

Thus this leads to a particular choice of an inner product inducing a metric in the subjects space. The mathematical development requires to define a PCA in a Sobolev space (Besse, 1980) or, more accurately, in a finite-dimension approximation of a Sobolev space.

Given that a particular spline interpolation is the best approximation according to a Sobolev norm, it is shown that the PCA performed with the induced metric amounts, as a consequence, to deal with spline interpolations of the curves (Besse and Ramsay, 1986).

Thus, in this particular use of a linear PCA the objects—the curves—are spline interpolated as opposed to the spline approximation of data transformations in the non-linear approaches.

Sections 2, 3, 4 and 5 briefly describe the convenient mathematical tools. Section 6 gives some results concerning the application to the linear PCA. Section 7 discusses the 'best' choice of a parameter defining the metric—that is the 'optimal' metric—in order to get a 'good' PCA leading to 'optimal' graphical displays. Section 8 deals with illustrative practical examples.

2. SOBOLEV INNER PRODUCT

Let $T=[0, 1]$ and $L^2(T)$ be the space of functions whose squares have a finite Lebesgue integral over T; $L^2(T)$ is an Hilbert space with the classical inner product:

$$(u, v)=\int_T u(t)v(t)\,dt.$$

A Sobolev space $H^m(T)\,(m\in\mathbf{N})$ is defined as a space of functions having absolutely continuous derivatives up to order $m-1$ and the mth order derivative in $L^2(T)$. If D^m denotes the mth order derivative operator:

$$H^m(T)=\{u\in C^{m-1}(T);\ D^m u\in L^2(T)\}.$$

Consider the linear differential operator:

$$L_\beta=\sum_{i=1}^{m}\beta_i D^i;\ \beta_i\in\mathbb{R},\qquad \beta_m\neq 0.$$

Ker L_β is the subspace of functions within $H^m(T)$ that verifies the homogeneous linear differential equation: $L_\beta u=0$; thus H^m is expressed as the direct sum:

$$H^m=\text{Ker } L_\beta\oplus\text{Ker } L_\beta^\perp$$

and Ker $L_\beta^\perp$ is an Hilbert space with the following inner product:

$$(u, v)_1=(L_\beta u, L_\beta v).$$

Furthermore, any function u in H^m can be represented by a single decomposition:

$$u=u_0+u_1,\quad u_0\in\text{Ker } L_\beta\quad\text{and}\quad u_1\in\text{Ker } L_\beta^\perp;$$

u_0 is called the 'trend' and u_1 the residual variation or 'noise'.

A classical way, to define an inner product $(.,.)_{H^m}$ for H^m, is to consider m bounded linear functions l_i such that:

$$\forall u \in \operatorname{Ker} L_\beta, \qquad \begin{array}{l} l_i(u)=0 \Rightarrow m=0. \\ i=1,\ldots,m \end{array}$$

Thus, $(.,.)_{H^m}$, $(.,.)_0$ and $(.,.)_1$ denote inner products for H^m, $\operatorname{Ker} L_\beta$ and $\operatorname{Ker} L_\beta^\perp$ respectively with:

$$(u,v)_0=\sum_{i=1}^{m} l_i(u)l_i(v) \quad \text{and}$$

$$(u,v)_{H^m}=(u,v)_0+(u,v)_1.$$

The functions l_i usually are some boundary conditions; for instance, if $m=1$:

$H^1(T)=\{u\in C^0(T);\ Du\in L^2(T)\}$,

$l_1(u)=u(0)$,

Ker D is spanned by the constant function and

$(u,v)_{H^1}=u(0)v(0)+(Du,Dv)$.

3. THE REPRODUCING KERNEL

3.1. Notation and basic properties

For any subset $T\in\mathbb{R}$ and Hilbert space H of real functions defined on T, a bivariate function mapping $T\times T$ onto $\mathbb{R}$ is called a reproducing kernel (r.k.) for H if $k(s,.)$ belongs to H for all s within T and if it verifies the basic reproducing equation:

$$\forall u\in H,\ \forall s\in T,\ (u,k(s,.))_H=u(s).$$

It works, for the H inner product, as a Dirac distribution applied to a continuous function or as M^{-1} in a finite-dimensional vector space with metric M. The proofs of the following properties can be found in Aronszajn (1950), Duc-Jacquet (1973) or Shapiro (1971).

(1) k is symmetric: $k(s,t)=k(t,s)$,
(2) k is positive, it means that the matrix K with elements $k(t_i,t_j)$ is positive semidefinite for any p-uple $(t_1,\ldots,t_p)$ of elements in T.
(3) k is unique for a given space H.
(4) The vector space spanned by $\{k(s,.),\ s\in T\}$ is dense in H.
(5) If H is a direct sum $H_0\oplus H_1$ of Hilbert spaces H_0 and H_1 with respective reproducing kernels k_0 and k_1 then:

$$k(s,t)=k_0(s,t)+k_1(s,t).$$

(6) The r.k. for a finite-dimensional Hilbert space H spanned by a basis $u_1, \ldots, u_p$ is given by

$$k(s, t) = \sum_{i=1}^{p} \sum_{j=1}^{p} b_{ij} u_i(s) u_j(t)$$

where b_{ij} is the ij^{th} element of the inverse of the matrix with elements $(u_i, u_j)_H$.

3.2. Some examples

For the above example H^1 of Sobolev space:

$$k_o(s, t) = 1,$$
$$k_1(s, t) = \min(s, t) \text{ and}$$
$$k(s, t) = 1 + \min(s, t)$$

are r.k. for Ker D, Ker $D^{\perp}$ and H^1 respectively with the inner product:

$$(u, v)_{H^1} = u(0)v(0) + (Du, Dv).$$

Of course, a much more general formulation of inner products and associated r.k. can be considered on a Sobolev space H^m. The inner products family:

$$(u, v)_{\alpha, \beta, l} = (1-\alpha)(u, v)_0 + \alpha(L_\beta u, L_\beta v)$$

with

$$\beta \in \mathbb{R}^m, \beta_m \neq 0, \alpha \in]0, 1[, l = \{l_i; \mathrm{i} - 1, \ldots, \mathrm{m}\}$$

satisfying (1), leads to equivalent norms in H^m. Since the r.k. calculus is not always easy in any case and such a complexity does not seem to be required in practice, only simple cases, such as the following ones, will be considered in this chapter:

(a) $$(\,.\,,\,.\,)_{H^m} = (1-\alpha)(\,.\,,\,.\,)_0 + \alpha(\,.\,,\,.\,)_1, \; m = 1 \text{ or } 2, \alpha \in]0, 1[.$$

Since the linearity of the inner product, the associated r.k. is given by:

$$k_\alpha(s, t) = (1-\alpha)^{-1} k_0(s, t) + \alpha^{-1} k_1(s, t).$$

Thus each component is balanced following the α value. The closer α is to zero, the more the inner product takes into account the 'trend'. On the other hand, the closer α is to one, the more this 'trend' is neglected and the 'residual' important.

(b) $$(u, v)_\beta = (L_\beta u, L_\beta v) \textit{ with}$$
$$L_\beta = (1-\beta)I + \beta D \quad \text{and} \quad \beta \in]0, 1[.$$

Then the r.k. becomes:

$$k_\beta(s,t)=[\beta(1-\beta)]^{-1}\exp\left(\frac{\beta-1}{\beta}s\right)\sinh\left(\frac{1-\beta}{\beta}t\right),\ t\leqslant s,$$

and since k_β is symmetric,

$$k_\beta(s,t)=k_\beta(t,s) \text{ when } t>s.$$

In this case, the closer β is to zero, the more the inner product will approach that of $L^2=H^0$. Thus β balances respective influence of absolute values and growths in the dissimilarities calculus between the curves.

General tools, using Green functions or cubic spline approximation, are described by Besse and Ramsay (1986) in order to compute the r.k. and some other applications.

4. SPLINE INTERPOLATION

The n functions to study are only known at p sampled points $\{t_1, \ldots, t_p\}$ or knots belonging to T. Assuming that the curves lie in a prior chosen space H^m, the problem is to seek some good representative functions in that space. A natural way is to choose interpolating functions at sampled points that minimize the H^m norm: if f belongs to H^m and W denotes the set of all functions of H^m which fit f at each t_j,

$$W=\{u\in H^m \mid u(t_j)=f(t_j);\ j=1,\ldots,p\},$$

the spline interpolation of f is the function h of W which minimizes:

$$\|h\|_m=\operatorname*{Min}_{u\in W}\|u\|_m.$$

h is the smoothest interpolation for this norm.

The definition of a r.k. permits a simple account to define general spline interpolation in the frame of a reproducing Hilbert space (see the above references). Thus h is the orthogonal projection of f on the p dimensions subspace S_p of H^m spanned by the r.k.:

$$S_p=\{k(t_j,.);\ j=1,\ldots,p\},$$

and the shape of the interpolator between the knots, which can be piecewise linear, exponential, polynomial, . . . , depends on the choice of the inner product in H^m. For simplicity, $\{t_1, \ldots, t_p\}$ is assumed to be Ker L_β-unisolvent. It means:

$$\text{if}\quad u\in \operatorname{Ker} L_\beta,\quad u(t_j)=0;\quad j=1,\ldots,p\quad \text{then}\quad u=0.$$

Thus, the sequence $\{k(t_j,.);\ j=1,\ldots,p\}$ defines a basis in S_p and the simple solution of (2) is obtained by solving the following linear system:

$$h(t_i)=\sum_{j=1}^{p} a_j k(t_i,t_j)=f(t_i);\quad i=1,\ldots,p.$$

Then, in the r.k. basis,

$$a=K^{-1}u,$$

where $a=[a_1, \ldots, a_p]'$, $u=[f(t_1), \ldots, f(t_p)]'$, and K is the positive definite matrix containing the values of the r.k. at the sampling knots:

$$[K]_{ij}=(k(t_i, .), \quad k(t_j, .))H^m=k(t_i, t_j).$$

Thus, the knowledge of the r.k. leads to a direct solution for the minimum norm interpolating splines.

5. PCA OF INTERPOLATED FUNCTIONS

Consider now the classical PCA context: the observations of a variable X at p times or at p values of a real argument $\{t_j; j=1, \ldots, p\}$ on n objects are organized in a n rows and p columns matrix also denoted by X; x_i denotes the column vector in $\mathbb{R}^p$ containing the sampled values of the ith object, h_i its interpolant.

5.1. Results in H^m, H_0, H_1

Besse (1980) proposes a simple way to compute the PCA of interpolants:

Theorem 1 *When K is positive definite the PCA of interpolants $h_i \in H^m$ is equivalent to classical PCA of matrix X in the metric K^{-1}.*

Proof: With the above notations the inner product between two interpolants is

$$\begin{aligned}(h_1, h_2)&=a_1'Ka_2\\ &=x_1'K^{-1}KK^{-1}x_2\\ &=x_1'K^{-1}x_2.\end{aligned}$$ ■

Besse and Ramsay (1986) consider two other forms of PCA by using semi-metrics in order to point out some specific aspects of the data:

—In H_0 the goal is to study only the trend whose shape is defined by the homogeneous linear differential equation $L_\beta u=0$,
—In H_1 the goal is to study any meaningful variation in the residuals that meet the boundary conditions; $l_i(u)=0$; $i=1, \ldots, m$. This is achieved by:

Theorem 2 *The PCA of interpolating spline functions in H_0 (resp. in H_1) is equivalent to the classical PCA of X with semi-metric $M_0=K^{-1}K_0K^{-1}$ (resp. $M_1=K^{-1}K_1K^{-1}$).*

Proof: As above,

$$(h_1, h_2)_1 = a_2' K_0 a_2$$
$$= x_1' K^{-1} K_0 K^{-1} x_2 \qquad \blacksquare$$

Besse and Ramsay (1986) fully discuss these three types of PCA (in H^m, H_0, H_1) completely and compare them with the classical PCA (in L^2) and with the RAO, 1964's approach where some components can be regarded as instrumental variables.

But, in this chapter, our purpose is different, it is not to discuss a model choice or the meaning of residuals. It focuses on the choice of the inner product parameters, that are defined as in Section 3.2, in order to get some 'good' PCA. Since this last objective seems to be very intuitive, we are going to try to define it more accurately starting from empirical considerations.

6. NOTION OF 'OPTIMAL' METRIC IN PCA

Trivially, we can say that a PCA is 'good' if it leads to interesting and reliable comments or interpretations based on a little number $q < p$ of the first principal components (p.c.) well differentiated from the other ones. This little number must not be too small, that is to say greater than one since in this case the first component often takes only one trivial effect into account: the size of some objects, the variance of one or two variables, the redundancy between some variables .... It is then clear that a PCA is not required. On the other hand, q must not be too large since it is difficult to interpret successfully more than four components and impossible to build one's own understanding of such a space.

Thus the problem deals clearly with a question which has been widely developed in the literature: how many p.c. is it necessary to retain? Since the functional data context offers a large variety of metrics, this last and important question is turned into a question of metric choice: what metric leads to a PCA with a few well differentiated principal components? Besse (1986, 1987) develops an analogous approach with another type of data requiring to be smoothed; the same kind of reasoning leads to the choice of the best smoothing parameter.

This strategy follows a similar pattern to those developed by Winsberg and Ramsay (1983) (cf. Chapter 6) or de Leeuw *et al.* (1981) (cf. Chapter 3). The objective is here to look for a metric, that is equivalent to a linear transformation of the data columns, instead of non-linear, monotone or not, transformations.

6.1. The number of principal components

Besse *et al.* (1987) deal with this problem in the framework of the fixed effect model. Assuming that the data are lying in a $q < p$ dimension subspace (of H^m), a Gauss–Markov argument shows that the metric to use in a PCA is given by the inverse of the residual or noise covariance matrix; that is the H_1 metric.

Furthermore they propose a criterion based on a quadratic error showing in some examples, that the best graphical displays are not necessarily obtained with the real number q of p.c. Intuitively, if the unknown error variance is greater than the last q-q' eigenvalues then just q' p.c. have to be retained.

Jolliffe (1986, Ch. 6) reviews some frequently used rules for deciding how many p.c. should be retained. He distinguishes between *ad hoc* rules, intuitively plausible, and rules based on formal reasoning. Some of them require distributional assumptions to test the equality of the last eigenvalues while some are distribution free (cross-validation and partial correlation).

Following Jolliffe who concludes, that '*the rules which have more sound statistical foundations seem, at present, to offer little advantage over the simpler*' and considering that the PCA optimization problem is not yet well formalized, we will only consider, in this chapter, three simple criteria:

*R*1: Any statistical package performing a PCA provides the *cumulative percentage of total variation* for each eigenvalue:

$$t_k = 100 \times \sum_{i=1}^{k} \lambda_k / \mathrm{tr}(VM)$$

where $\{\lambda_i : i = 1, \ldots, p\}$ denote the eigenvalues of VM. Then a very popular rule is as follows:

Choose a cut-off t^* depending on p and n somewhere between 70 and 90 per cent and retain q p.c., where q is the smallest integer k for which $t_k > t^*$.

*R*2: The *scree graph* (Cattell, 1966) and the *log-eigenvalue* (*l.e.v.*) *diagram* (Craddock and Flood, 1969) are plots of λ_k and $\log \lambda_k$ respectively against k.

With the scree graph the user must decide at which value of $k(=q)$ the slopes of lines joining the plotted points are 'steep' to the left of the kth and 'not steep' to the right. In other words, the rule is to look for the point beyond which the scree graph almost defines a straight line. The rank of the first point on the straight line gives the q value.

The l.e.v. diagram developed in the field of meteorology seems to be more adapted when data are functions. Craddock and Flood (1969) argues that eigenvalues corresponding to 'noise' should decay in a geometric progression and thus appear as a straight line on the l.e.v. diagram. It leads to the same decision rule but not to the same results.

*R*3: A third rule can be defined on the base of *bootstrap* (Daudin, Duby and Trecourts, 1987) or of *perturbation theory* (Besse *et al.*, 1987) considerations. We say that a PCA is reliable, or stable, if the space spanned by the q first p.c. is protected against small data fluctuations. It is achieved if the difference $\lambda_q - \lambda_{q+1}$ is large enough in order to prevent any switch between the qth and $q+1^{\mathrm{th}}$

eigendirections. Perturbation theory requires to assume that the noise variance is smaller than $\lambda_q - \lambda_{q+1}$.

This leads to fix the cut-off only between well differentiated eigenvalues; q is chosen such that both t_q and $\lambda_q - \lambda_{q+1}$ are 'large enough'.

To sum up, R1 is based on the λ_k values, R2 on the $\lambda_k - \lambda_{k+1}$ values and both are mixed in R3. Of course these criteria are largely subjective and, in general, do not lead to the same value for q. Note that R3 suggests that, in some cases, the q value induced by R2 must be replaced by $q-1$: it seems better to cut off just before a 'straight line' rather than at its beginning.

6.2. A 'good' PCA

Consider the scree graphs and l.e.v. diagrams in Figure 4.1 below. They have been obtained by performing PCA of data described by Besse (1986) with different values of a smoothing parameter.

For $a=0.10$, R1 and R3 indicate that only one component should be retained. This is a typical example of a trivial dominant first component. The scree graph (R2) leads to choose $q=1$ or 2 and the l.e.v. diagram rather gives $q=6$ or 7!

For $a=0.90$, R1 indicates a large q value (6 or 7) since the data seems very noisy and without any reliable structure (Deville, 1977): the whole of the l.e.v. diagram fits approximatively a straight line.

For $a=0.30$, all rules agree with a q value between 3 and 4. In this case the decision is easy since both the scree graph and the l.e.v. diagram show a fairly sharp 'elbow'. The cut-off point is chosen just after the change in slope.

This example leads to propose the following strategy:

When noise variance and real dimension q of the data are both unknown—the usual case in practice—and if a family of metrics can be used to perform

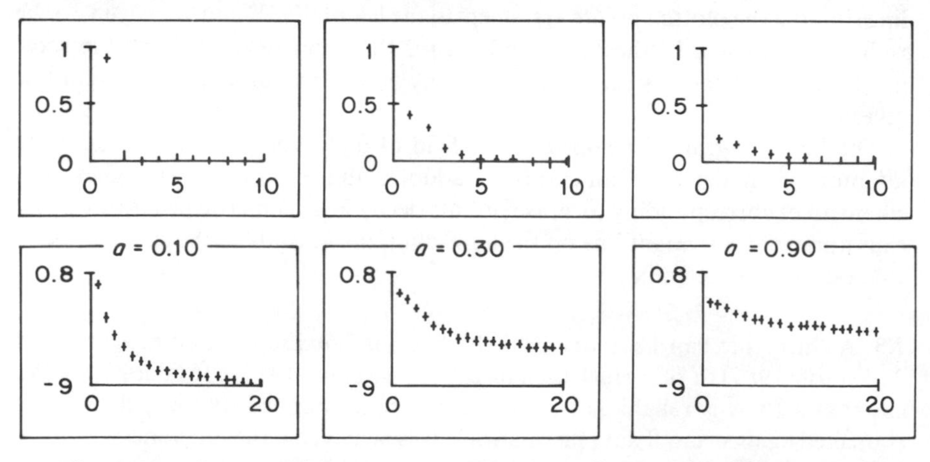

Figure 4.1. Scree graphs and corresponding l.e.v. diagrams for different metrics

PCA—for instance H^m for longitudinal data—a good PCA is obtained with the optimal metric providing the sharpest elbow scree graph. In other words the PCA leading to the 'clearest' cut-off choice.

To evaluate its sharpness, we consider an 'objective' criterion dealing with the second order differences of eigenvalues:

$$d^2 = \frac{\sup_{i=1}^{p-2} (2\lambda_{i+1} - \lambda_i - \lambda_{i+2})_+}{\sup_{i=1}^{p-2} (\lambda_i - 2\lambda_{i+1} + \lambda_{i+2})}$$

where $$(f)_+ = \begin{cases} f \text{ if } f \geqslant 0 \\ 0 \text{ if } f \leqslant 0 \end{cases}$$

Figure 4.2 displays d^2 values against different values of the parameter. A maximum occurs for a value close to 0.4 and both the scree graphs and the l.e.v.

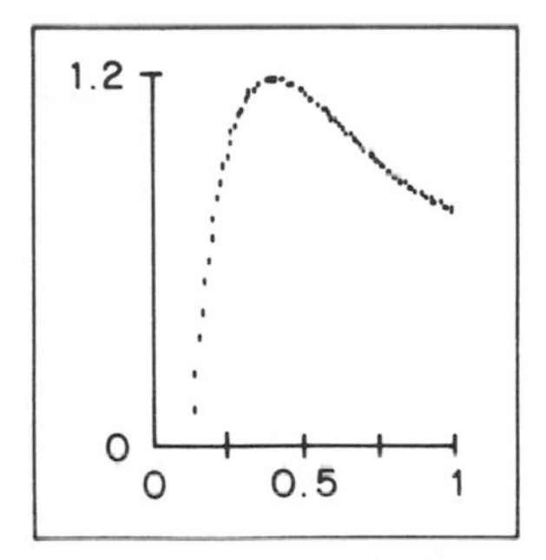

Figure 4.2. d^2 values when the smoothing parameter increases

diagrams (Figure 4.3) show how the elbow appears and then vanishes in the noise when the smoothing parameter increases.

Any convenient optimization algorithm could provide the best parameter value but, in this heuristical approach, it seems more important to study the behaviour of scree graphs, l.e.v. diagrams and any kind of criterion as d^2 on further examples first. After that it may be possible to formalize this work better.

7. SOME ILLUSTRATIVE EXAMPLES

7.1. Flow/volume curves

These curves are used to analyse obstructive syndromes of breathing. The first curves—flow as a function of time—are obtained by inviting the patients to blow through a tube equipped with a flow captor. The second ones—flow as a function

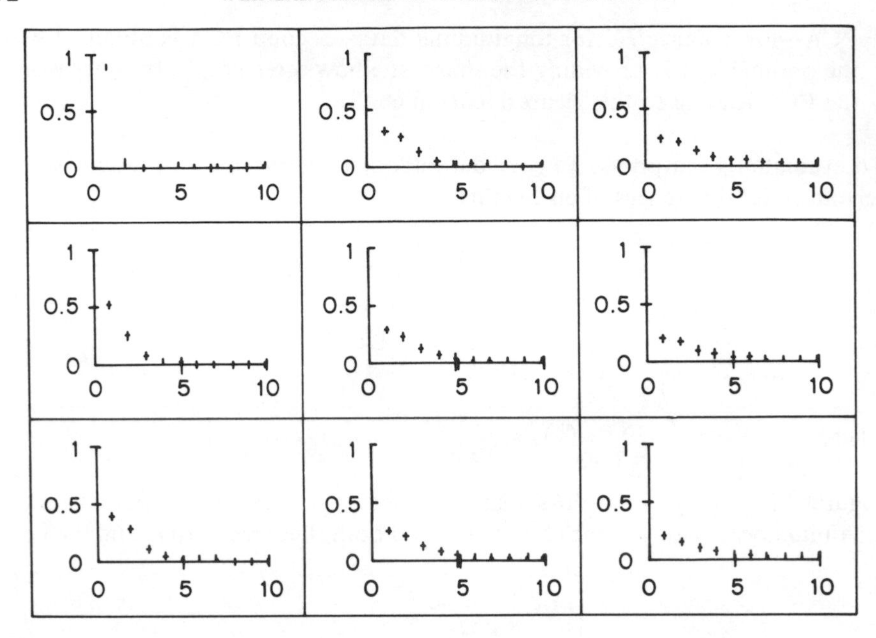

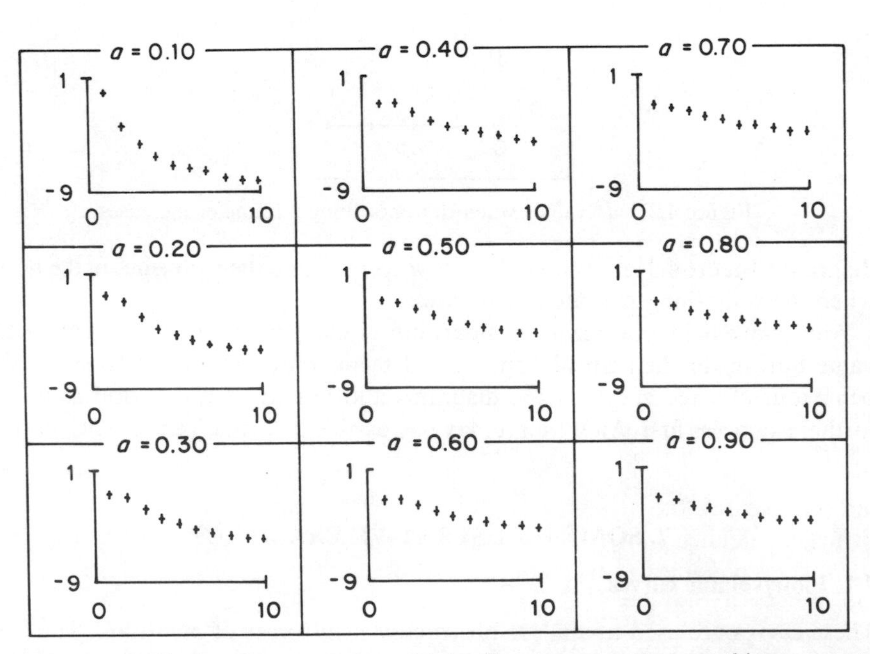

Figure 4.3. Evolution of scree graphs and l.e.v. diagrams when the smoothing parameter increases

of volume—are obtained by integration. Finally the x-axis scale is changed to measure a percentage of the total volume expired. (We are grateful to Drs Pecoul and Tap from Professor Besombes's service at the Centre Hospitalier Universitaire de Toulouse who collected the data.) Physicians usually deal with the latter curves but as the 'peak flow' strongly depends on the height and weight of the subjects, the curves have been 'normalized': the 'peak flow' is taken equal to 1 for each curve.

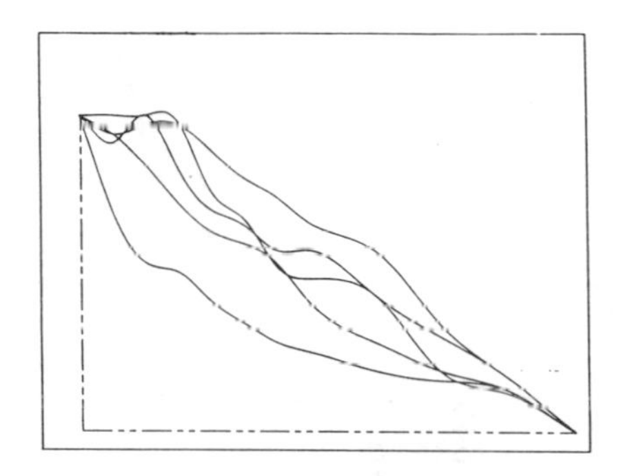

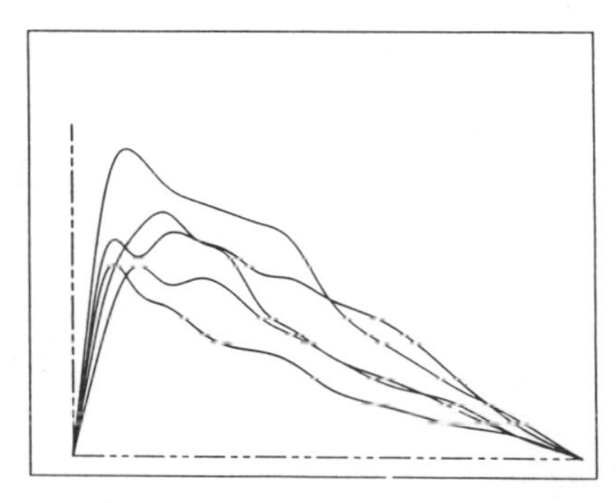

Figure 4.4. Cubic spline interpolation of some flow/volume and 'normalized' flow/volume curves

The measures have been made for $n=40$ boys aged 10 to 14 and suffering from respiratory diseases. Since the curves are directly displayed on a plotter, a discretization is required before computing any PCA; we chose to use $p=7$ equidistant knots.

The first component of classical PCA (Figure 4.5) shows how serious the illness is at bronchioles level and is connected to overall shape of the curves. The second axis discriminates subjects whose curves show a lot of perturbations. It seems to indicate that the troubles concern the bronchiolar level.

Now we consider a new metric induced by the following inner product:

$$(u, v)_1 = (L_u, L_v) \qquad \text{where } L=(1-\beta)I+\beta D,\ \beta \in]0, 1[;$$

it leads to use the r.k. described in Section 3.2(b) to build in the matrix defining the metric. Then several PCA are computed for different values of β between 0 and 1 to balance the respective influence of curves or of their derivatives in the distances calculus.

Thus it is possible to consider the behaviour of scree graphs, of l.e.v. diagrams and of the criterion d^2:

Figure 4.6 shows that a clear elbow appears on both scree graphs and l.e.v. diagrams when β belongs to the subset [0.60, 0.70]; here in this case all the cut-off rules lead to take $q=2$. More accurately, Figure 4.7 plots d^2 values against β. It

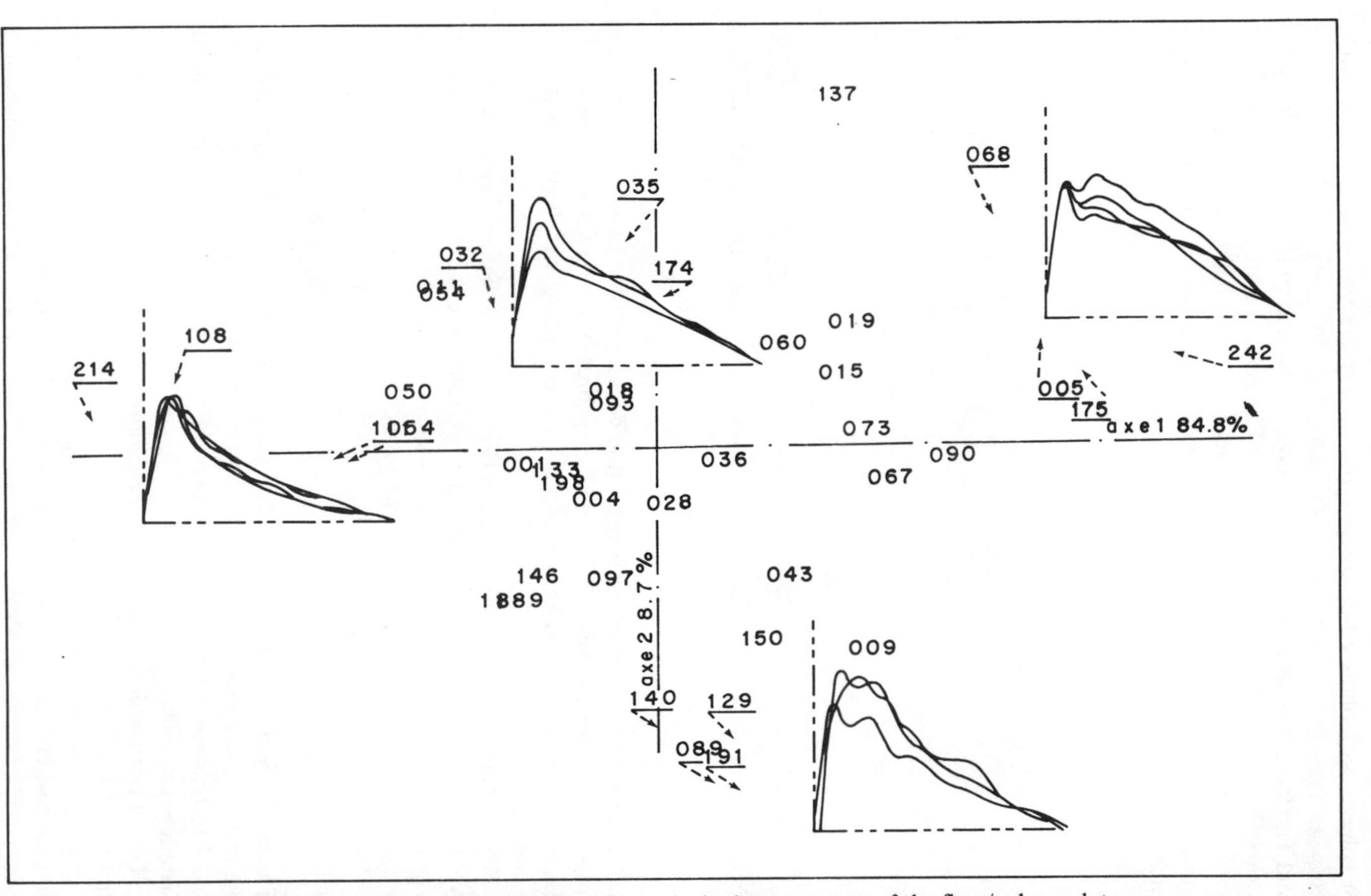

Figure 4.5. Scatter plot in the two first principal components of the flow/volume data

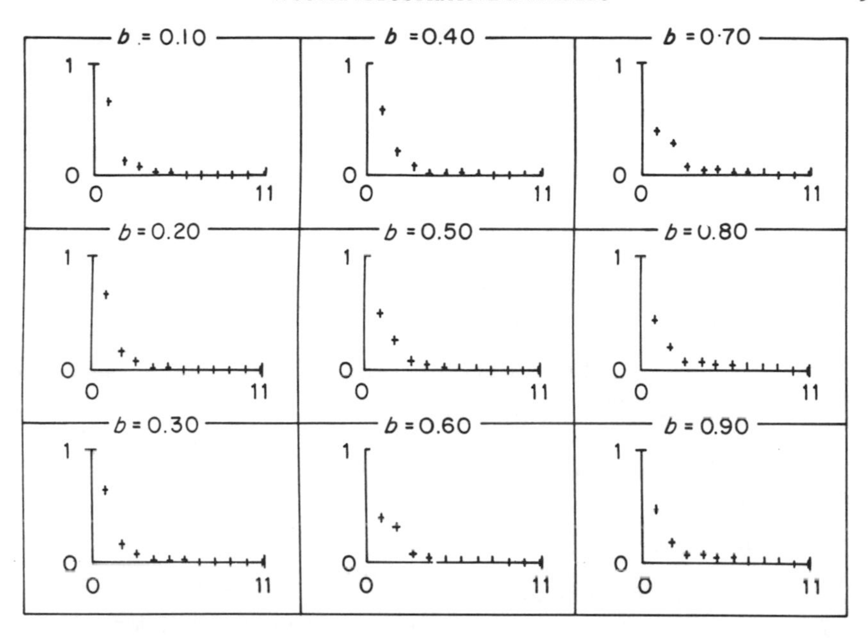

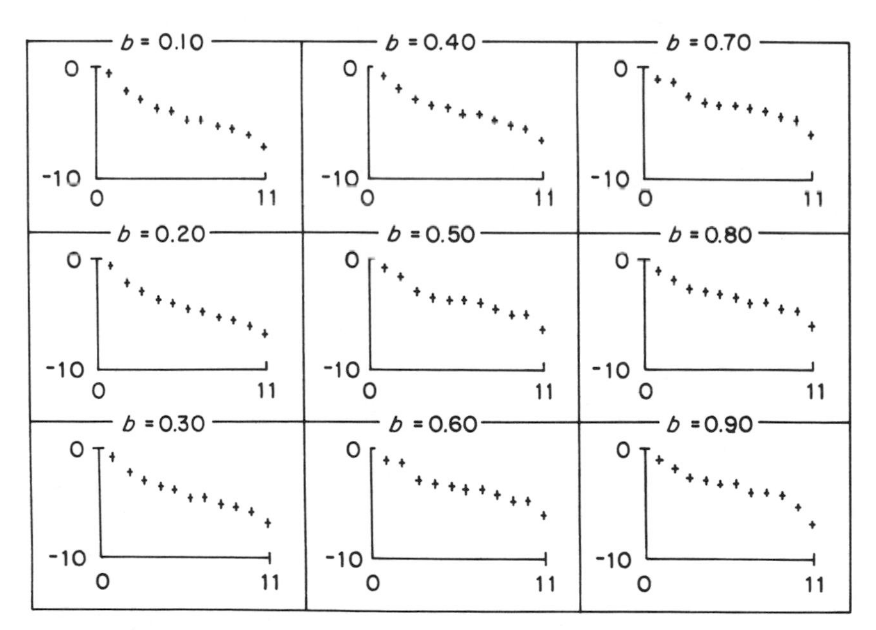

Figure 4.6. Evolution of scree graphs and l.e.v. diagrams when β increases

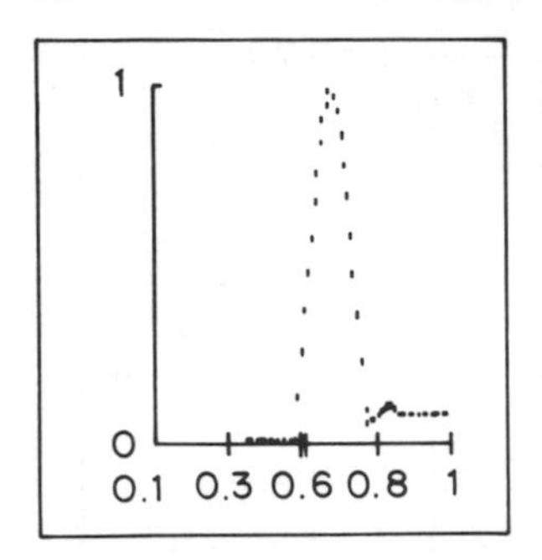

Figure 4.7. d^2 values when β increases from 0.10 to 0.99 by 0.01

It shows that the optimal metric is obtained for β to 0.64 (up to one hundred different values of β were tried).

Clearly, the results above lead to retain the metric associated with $\beta=0.64$ and display the data in a space of dimension $q=2$.

7.2. Fecundity rates

The following example is similarly treated. It deals with the fecundity rate of $n=27$ industrialized countries. It is the ratio of the number of children born in a year over the number of women who can give birth. The values found in Munoz-Perez (1982) pertain to $p=9$ years (1971 to 1979). Some of them are displayed in Figure 4.8. Two families of metrics have been used in order to find an optimal one:

(a) The first one is induced by the following inner product:

$$(u, v)_{M'}=(\alpha-1)u(0)v(0)+\alpha(Du, Dv), \qquad \alpha\in]0, 1[$$

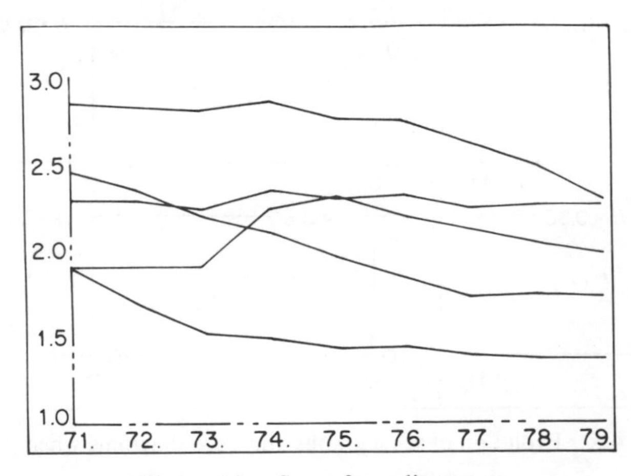

Figure 4.8. Some fecundity rates

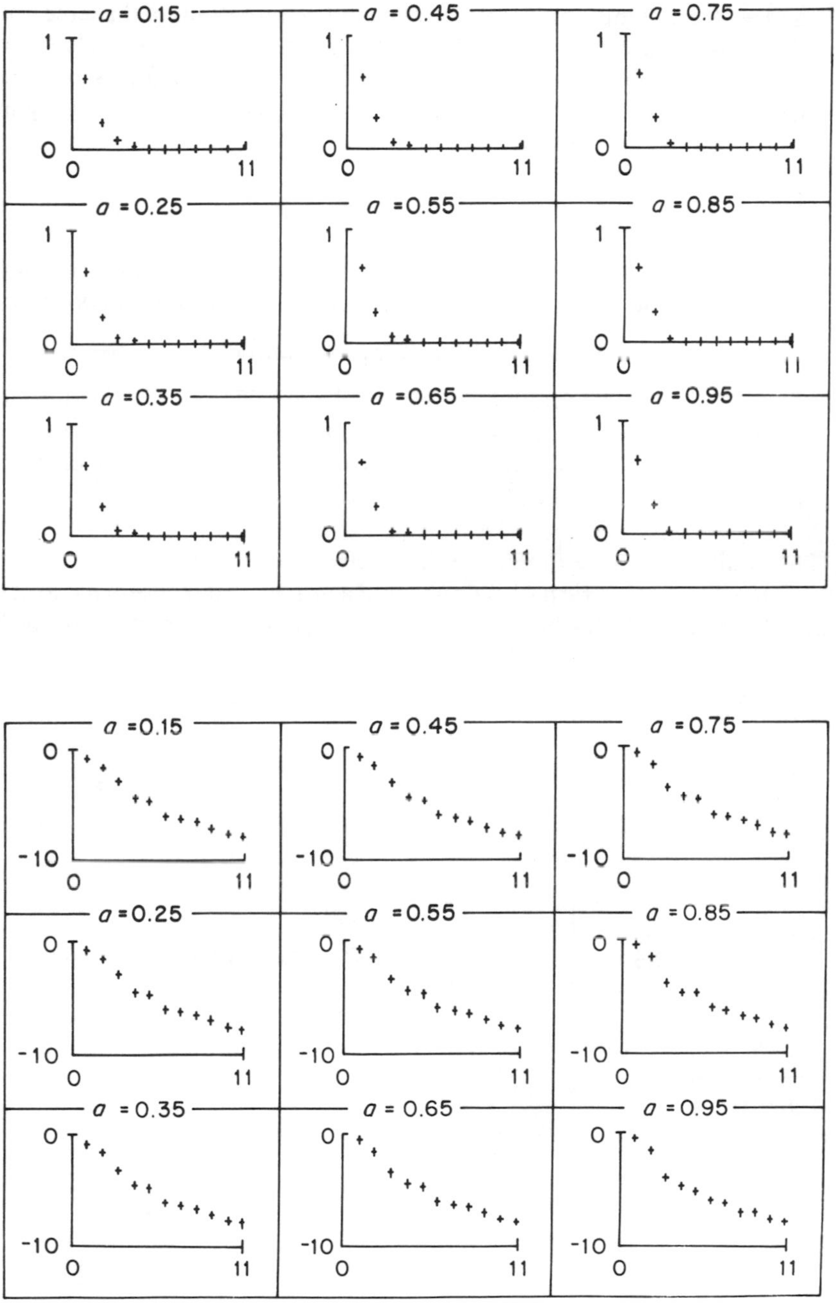

a = 0.15
a = 0.45
a = 0.75
a = 0.25
a = 0.55
a = 0.85
a = 0.35
a = 0.65
a = 0.95

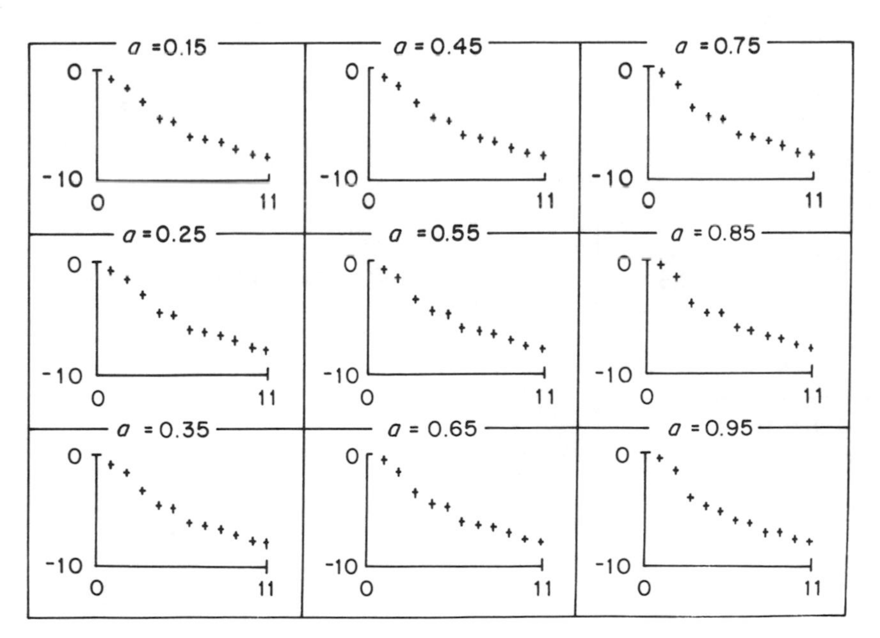

Figure 4.9. Scree graphs and l.e.v. diagrams

and we are looking for an optimal α value. The evolutions of the scree graphs do not bring any help. However the l.e.v. diagrams, with an α value around 0.85, indicate an interesting choice for $q=2$ (see Figure 4.9). This is corroborated by the d^2 plot (Figure 4.10) showing an optimal but very small value (0.0021) for $\alpha \approx 0.86$.

(b) The second family of metrics is induced by:

$$(u, v)=(Lu, Lv) \qquad \text{where } L=(\beta-1)I+\beta D, \ \beta \in]0, 1[.$$

The parameter β balances between the influence of a constant component and the variation of the curves. In this case the d^2 plot shows two maxima:

$$d^2(\beta_1)<d^2(\beta_2) \qquad \text{with } \beta_1 \approx 0.3 \text{ and } \beta_2 \approx 0.6.$$

For $\beta=\beta_1$, the scree graph and the l.e.v. diagram do not seem to agree on a same cut-off point (Figure 4.12). The first leads to $q=2$ while the second to $q=4$. For $\beta=\beta_2$ the $d^2(\beta_2)$ value indicates a sharper elbow. Correspondingly, the l.e.v. and scree plots lead to choose the same cut-off point at $q=3$.

The results above lead to retain either metric a with $\alpha=0.86$, and a display in dimension $q=2$, or metric b with $\beta \approx 0.6$ and a display in dimension $q=3$. Metric b might be selected on the basis of the d^2 values. To illustrate partly how the final

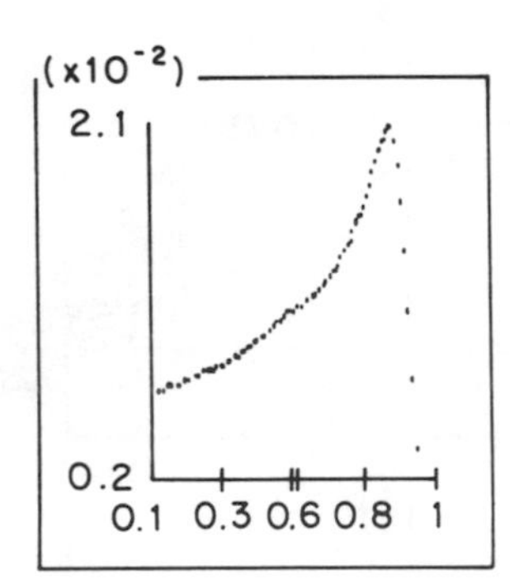

Figure 4.10. d^2 values when α increases from 0.10 to 0.99 by 0.01

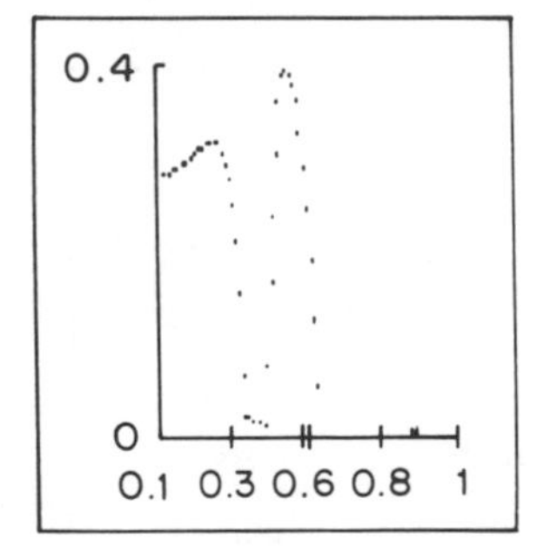

Figure 4.11. d^2 values when β increases from 0.10 to 0.99 by 0.01

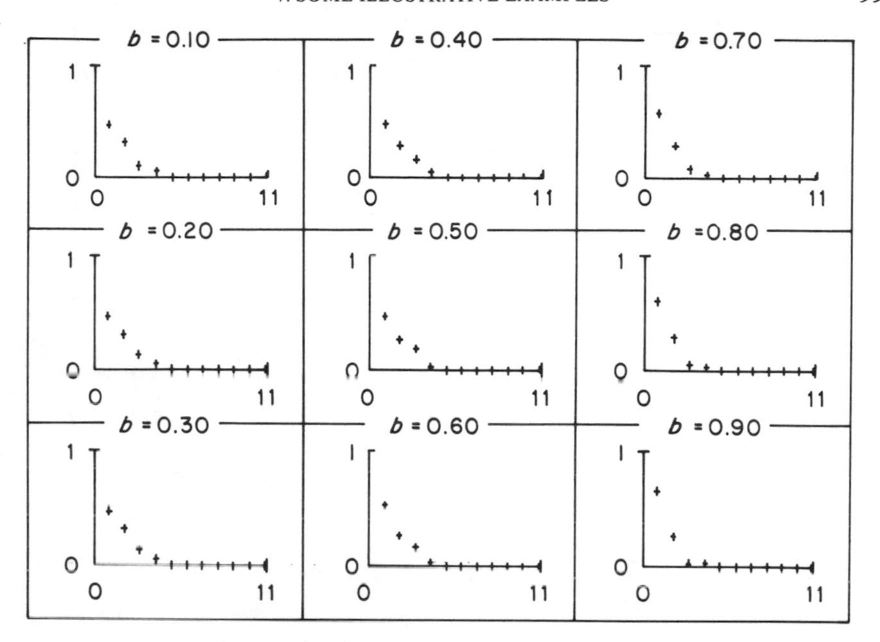

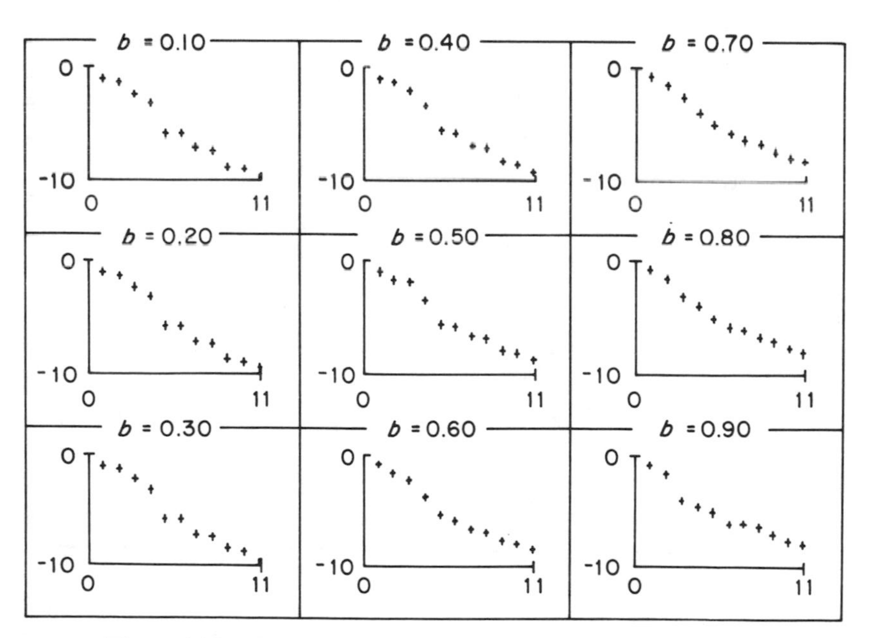

Figure 4.12. Scree graphs and l.e.v. diagrams when β increases

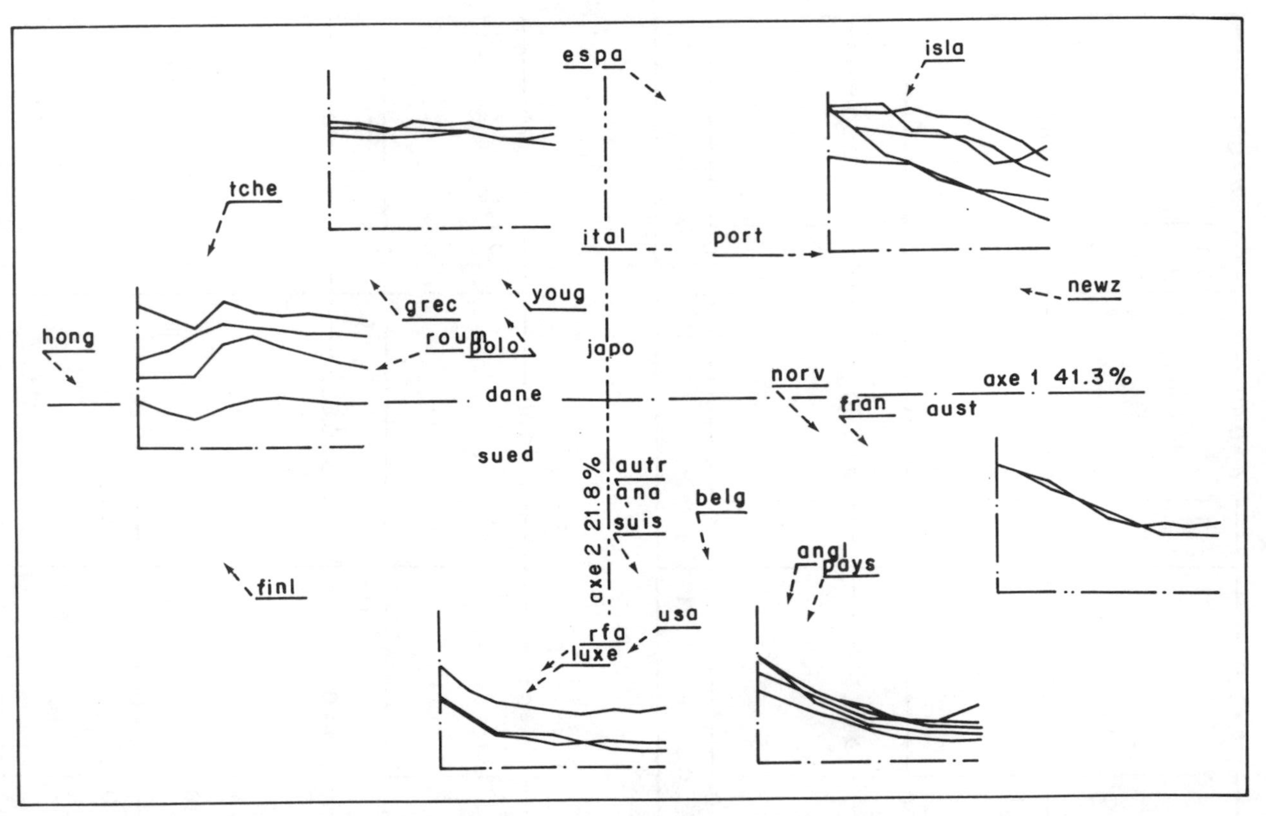

Figure 4.13. Scatter plot in the two first principal components of fecundity rates data

display of the initial data (fecundity rate over time) might look, their scatter-plot in the first two dimensions is given in Figure 4.13. The display opposes the countries, whose rate decreases quickly, to Eastern countries whose rate increased in 1973–74. Countries with an almost constant rate are situated in the neighbourhood of the origin. In the upper part of the diagram are countries whose rate steadily decreases during the period; they are opposed to countries whose rate becomes constant or even begins to increase again in some cases. This second axis discriminates high- and low-rate countries in a certain way because high rates can decrease quickly whereas the low rates seem to come to a standstill after a fall previous to the considered period.

Component and Correspondence Analysis
Edited by J.L.A. van Rijckevorsel and J. de Leeuw

Chapter 5

On Probability Coding

J.-F. Martin
INSA, Centre de Mathématiques, Villeurbanne, France

1. INTRODUCTION

The concept of P-coding (or fuzzy coding) arises more particularly in the discretization of continuous real random variables (r.r.v.) on the intervals of a partition of R (set of real numbers): it consists of generalizing the Dirac coding (which attributes an observation to an interval) by associating eventually an observation to several intervals, with weights related to these intervals. See also Chapter 2.

In this chapter, we shall give a definition of P-coding, obtained by a probabilistic interpretation of this concept: to code an observation $x = X(\omega)$ of a random variable (r.v.) X in ω means to associate to x a distribution of probability P_x.

If we intend to keep the 'primary' data x, without introducing any 'fuzziness', we can use the Dirac coding $P_x = \delta_x$ (δ_x being the Dirac measure). If, for some reason or other (random error, smoothing of data, . . .) we want to use some prior information (objective or subjective), we can use a P-coding. We shall study some examples of this situation.

The probabilistic interpretation of P-coding enables us to enlarge its applicabilities. We shall study more particularly:

(1) The coding of categorical ordered r.v. generated by a continuous r.r.v.: for example the categorical r.v. is x = 'do you watch TV (a little, moderately or much)', observed on a given population, generated by the r.r.v. Y = 'the time spent watching TV every week'. We can use a P-coding of X to give an account of the 'fuzziness' between the observed variable X and the underlying phenomenon Y.
(2) The smoothing of data with a P-coding (obtained by a convolution). This

coding gives an interesting interpretation of the kernel method in the density estimation for continuous or categorical data.

(3) The χ^2-test of coded data: in order to test the fit between the distribution of a continuous r.r.v. X and a given distribution, we must, in the χ^2-procedure, discretize the distributions on intervals. We can use a P-coding for this discretization and obtain a test less sensitive to the fluctuations of the model (choice of intervals, sample fluctuations, . . .).

(4) The PCA of a coded sample: classical PCA consists of the geometrical study of a cloud of n points. If we code the data, we must study a cloud of n probability distributions. We shall analyse this situation and study, with an example, the effect of P-coding upon the results of PCA. See also Chapters 4 and 6.

2. NOTATION AND DEFINITIONS

r.v.	random variable
r.r.v.	real random variable
PCA	principal component analysis
$B(I)$	σ-algebra of Borel sets of I
$\sigma(A_1, \ldots, A_q)$	σ-algebra generated by $\{A_1, \ldots, A_q\}$
1_F	indicator function of F: $1_F(x)=1$ if $x \in F$; $1_F(x)=0$ otherwise
R	set of real numbers
δ_x	Dirac measure on a σ-algebra B: if $F \in B$ $\delta_x(F)=1_F(x)$
B_R	σ-algebra of Borel

Let Y a r.v. defined on the space (Ω, a, μ), with values in the space (E_1, F). A common statistical problem is that, in order to get information on the variable Y, we must observe another variable X (more precisely, observe a sample $(X_1, \ldots, X_n)$ of an independent r.v. having the same distribution as X), while X might be but an 'approximation' of Y. For example X is a categorical r.v. originating with a set of questions meant to give an account of a phenomenon represented by Y, if X is a discretization of the continuous r.r.v. Y on intervals of a partition of R (histogram), or also X is the r.v. Y tainted with a measurement or method error, etc. . . .

When the statistician decides not to deal with the subjacent phenomenon, concretized by Y, but only with his data $X_1(\omega), \ldots, X_n(\omega)$, he uses the usual statistical tools: descriptive statistics, data analysis, inferential statistics, which led him to draw conclusions on the distribution of X.

However he must sometimes draw from these data $(X_1(\omega), \ldots, X_n(\omega))$ some information about the variable Y. Let, for example, the variable Y measured in ω by a device giving the value $X(\omega)$ which is a function of the theoretical value $Y(\omega)$ and of the value $E(\omega)$ of a random error: objective considerations (knowledge of the device) and/or subjective ones (error model) can often help us to draw from the observation of X some information about Y. We are interested in this kind of problem: how to formalize the 'randomness', as an object which we want to study, around the observations that are meant to represent a phenomenon through errors, reductions, etc.

The general idea is derived from the following intuitive approach: let us suppose that, for some reason, we must discretize a continuous r.r.v. Z on the intervals of a finite partition of R. To an observation z of Z we associate the interval which contains z. But if for example z is near the limit common to two intervals, a small change of the model (sample fluctuation, slight change of the limits between the intervals, dubiousness of measurement) can lead to big changes in the distribution of the observations in the intervals.

We can then just like in Chapter 2 consider attributing to this observation weights related to each interval, these weights being two positive numbers, the sum of which is 1: we have associated to the observation z a distribution of probability P_z on R provided with the finite σ-algebra generated by the intervals.

This intuitive idea can be formalized in the following way. As opposed to the situation in which only the data $(X_1(\omega), \ldots, X_n(\omega))$ are given an account, we are going to associate to each observation a distribution of probability on the measurable space (E_1, F) in which Y gets its values.

This distribution is supposed to be the mathematical model of the 'fuzziness' between Y and X. To a sample of observations $(X_1(\omega), \ldots, X_n(\omega))$ will then be associated a coded sample $(P_{x1(\omega)}, \ldots, P_{xn(\omega)})$ of distributions of probability on (E_1, F).

2.1. Definitions

(a) Let (E, E) and (E_1, F) be two measurable spaces. A P-coding (on (E, E), associated to F) is a transition distribution of probability P on $(E, \mathrm{E}) \times \mathrm{F}$:

$$P: (x, F) \rightarrow P_x(F)$$

i.e. $(\forall F \in \mathrm{F})$, $P(F)$ is a measurable function and $(\forall x \in E) P_x$ is a distribution of probability on (E_1, F).

(b) Whatever F in F, the measurable function ϕ_F mapping (E, E) [0, 1] into B [0, 1] defined by:

$$\phi_F: x \rightarrow \phi_F(x) = P_x(F)$$

will be called a coding function (associated to F).

—To code an observation x of X in ω means to associated to $x = X(\omega)$ the distribution of probability P_x

—To code the r.v. X means to associate to X the P-coding:

$$P_x: \omega \rightarrow P_x(\omega).$$

Thus, to the primary data $(x_1, \ldots, x_n)$ are associated the coded data $(P_{x1}, \ldots, P_{xn})$.

Note that in the particular case in which $(E, \mathrm{E})=(E_1, \mathrm{F})$, the Dirac coding $P_x=\delta_x$(δ_x being the Dirac measure: $\delta_x(\mathrm{F})=1_F(x)$, indicator function of F) amounts to not changing the primary data, without introducing any 'fuzziness'.

3. P-CODING AND DISCRETIZATION OF *A*

This section completes the set of coding functions in Chapter 2. Given X a r.r.v. and $(A_1, A_2, \ldots, A_q)$ a partition of R into intervals, we can associate to an observation $x=X(\omega)$ weights

$$\phi_1(x), \ldots, \phi_q(x) \text{ related to } A_1, \ldots, A_q, \text{ with } \phi_j(x)\in[0, 1] \text{ and } \sum_j \phi_j=1$$

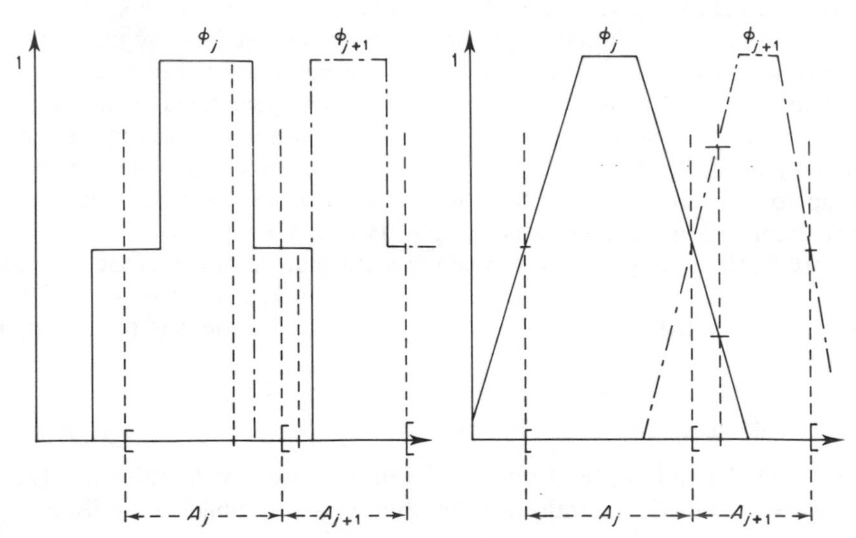

Figure 5.1. Two examples of coding functions

x_1 is coded by P_{x_1}: $\mathrm{P}_{x_1}(\mathrm{A}_j)=1$, $\mathrm{P}_{x_1}(\mathrm{A}_k)=0$ if $k\neq j$;

x_2 is coded by P_{x_2}: $P_{x_2}(A_j)=P_{x_2}(A_{j+1})=1/2$; $P_{x2}(A_k)=0$ if $k\neq j$ and $k\neq j+1$;

x_3 is coded by P_{x_3}: $\mathrm{P}_{x_3}(\mathrm{A}_j)=1/4$; $\mathrm{P}_{x_3}(\mathrm{A}_{j+1})=3/4$; $\mathrm{P}_{x_3}(\mathrm{A}_k)=0$ if $k\neq j$ and $k\neq j+1$.

If x is in the centre of A_j, we attribute x to A_j with the weight 1; but if x approaches the border between A_j and A_{j+1}, we attribute two weights to x, corresponding to the two intervals. $x=X(\omega)$ is then coded by the distribution of probability $P_{X(\omega)}$ on $(R, B_q=\sigma(A_1, \ldots, A_q))$ given by $P_X(A_j)=\phi_j(X)$. P_x is a P-coding of X, where the function ϕ_j is equal to the coding function. If $\phi_j=1_{A_j}$ we deal with crisp coding.

Let $(X_1, \ldots, X_n)$ be an n-sample of X.

—if $P_X = \delta_X$ (Dirac measure on $(R; \sigma(A_1, \ldots, A_q))$)

The coded sample is $(\delta_{X1}, \ldots, \delta_{X_n})$ and its empirical mean is the random measure $\delta = n^{-1} \sum_k \delta_{Xk}$. We see that:

$$\delta(A_j) = n^{-1} \sum_k \delta_{Xk}(A_j) = n^{-1} \sum_k 1_{A_j}(X_k) = n^{-1}$$

The observed relative frequency of A_j, $\delta(A_j)$ converges asymptotically to $E(1_{A_j}(X)) = \mu_X(A_j)$ (= probability of A_j), according to the strong law of large numbers (μ_X is the distribution of X).

—if P_X is any P-coding

The empirical mean of the coded sample $(P_{X1}, \ldots, P_{Xn})$ is now the random measure $v = n^{-1} \sum_k P_{Xk}$

$$v(A_j) = n^{-1} \sum_k P_{Xk}(A_j) = n^{-1} \sum_k \phi_j(X_k) = n^{-1},$$

which is equal to the sum of weights of observations to A_j.
Now, $v(A_j)$ (the coded observed relative frequency of A_j) converges to

$$E(\phi_j(X) = \int_R \phi_j(X)\, d\mu_X(x) \text{ (= the coded probability of } A_j).$$

4. CONDITIONAL CODING OF X IN RELATION TO Y

Let us go back to the case given in the introduction: X and Y are two r.v. defined on (Ω, a, μ), into (E, E) and E_1, F). Y is the r.v. which we would like to study and X the r.v. which we can observe. We would like, from observations of X, to get some information on Y. A reasonable idea is to associate to an observation x the conditional distribution of Y on x:

$$P_X = \mu_Y^{X=x}.$$

Of course, this approach remains theoretical in so far as the conditional distributions are usually not known. We shall then either have to estimate these distributions from a test sample on which X and Y will be known, or to approximate them by a subjective model (Martin, 1980). We can observe that the empirical mean of a coded sample of X:

$$v = n^{-1} \sum_k \mu_Y^{X=x}$$

converges asymptotically to $E(\mu_Y^X)=\mu_Y$, a distribution of the 'subjacent phenomenon' Y.

Example 1

$$E=E_1=R;\ E=F=B(R);\ Y=X+\varepsilon$$

Y is the sum of the observed variable X and with random error ε. If we suppose that X and ε are independent variables and that the distribution of ε is μ_ε, we can code X with the P-coding:

$$P_x=\mu_{Y^{X=x}}=\delta_x * \mu_\varepsilon.$$

Then we 'smooth' each observation with the distribution μ_ε. (cf. the second example of Section 6).

Example 2

X is the categorical r.v. 'do you watch TV (Moderately (L) or much (H))? and Y is the r.r.v. 'the time spent watching TV every week'. Let us suppose that we observe X and Y:

X	L	L	H	L	H	H	H	L	H	L
Y (hours)	1	3	8	7	10	5	4	3	10	1

Let us now search a number α such that the discretization of Y on the intervals $[0, \alpha]$ and $]\alpha, +\infty[$ will be the nearest to X. We can select $\alpha=3.5$, for example, with only one misclassification.
Let Z be the categorical variable $Z=$'L' if $Y\leq 3.5$ and $Z=$'H' if $Y>3.5$.
In order to give an account of the difference between X and Y, we can use the P-coding $P_x=\mu_Z^{X=x}$, estimated as follows: the answer 'moderately' ($X=$'L') will be coded:

$Y\leq 3.5$ with the probability 4/5

$Y>3.5$ with the probability 1/5

and the 'much' ($X=$'H') will be coded:

$Y>3$ with probability 1.

5. P-CODING AND THE KERNEL METHOD OF DENSITY ESTIMATION

Let X be a r.v.v. defined on (Ω, a, μ) for which we want to estimate the distribution μ_X, distribution of X. For that purpose, we have got the observation $(X_1(\omega), \ldots, X_n(\omega)) = (x_1, \ldots, x_n)$ of an n-sample of X.

We can then estimate the distribution function of X with the help of the empiric distribution function of the sample:

$$F_n(\omega; x) = n^{-1} \sum_k 1_{]-\infty, x]}(X_k(\omega))$$

which is the distribution function of the measure of probability:

$$\mu_n(\omega) = n^{-1} \sum_k \delta_{X_k}(\omega).$$

$\mu_n(\omega)$ is the empiric mean of the sample coded by the Dirac coding δ_Y:

$$(\delta_{X_1(\omega)}, \ldots, \delta_{X_n(\omega)}).$$

This approach can be justified by the fact that when n tends to infinity, $\mu_n(\omega; \mathrm{F})$ converges asymptotically to $E(\delta_x(F)) = E(1_F(X)) = \mu_X(F)$, according to the strong law of large numbers.

Then $\mu_n(\omega)$ does represent an approximation of the distribution μ_X of X. Let us suppose now that, for example to give an account of the sample fluctuations, we smooth each observation x_k according to a distribution π_n, around x_k. This amounts to coding each observation x_k with the distribution $P_{xk} = \pi_n * \delta_{xk}$. If this approach is meant to give an account of the sample fluctuations, perceptible when n is small, imperceptible when n is large, it is normal to assume that π_n converges to δ_0. Then as the sample size increases the less we smooth observations.

The coded sample is then $(\pi_n * \delta_{X_k(\omega)})_{k=1, \ldots, n}$ and its empirical mean is the distribution:

$$\mu_n(\omega) = n^{-1} \sum_k \pi_n * \delta_{X_k(\omega)}$$

having, for the Borel set F, the following value:

$$\mu_n(\omega; F) = n^{-1} \sum_k \pi_n(F - X_k(\omega)) = n^{-1} \sum_k \phi_f^n(X_k(\omega)),$$

which is the mean of the weights $\pi_n(F - x_k)$ of the observation x_k related to F, while ϕ_F^n is equal to the coded function associated to the coding $P_X^n = \pi_n * \delta_X$.

We can proceed in this way: let K be a kernel function (a positive, bounded and symmetric function of $L_1(\lambda)$, λ being the Lebesgue measure on R, while $\lim_\infty xK(x) = 0$ (cf. Chapter 4) with $\int_R K_X \, d\lambda = 1$. Let $(h(n))_N$ be a sequence of positive numbers with $\lim_\infty h(n) = 0$. We can easily prove that:

$$K_n : x \to h(n)^{-1} K(x) h(n))^{-1} \text{ has got the same properties as } K.$$

This kernel K is equal to the density of the probability distribution of π_n on R.

If we use this distribution π_n to code the r.v. X with $P_X = \pi_n * \delta_X$, we observe that the empirical mean of the coded sample is an absolutely continuous distribution with respect to λ. The density of λ is equal to:

$$f_n(\omega; x) = n^{-1} \sum_k K_n(x - X_k(\omega)) = (nh(n))^{-1} \sum_k K(x - X_k(\omega))\,(h(n))^{-1},$$

which is nothing else than the density given by the kernel method of density estimation. See also Bowman (1980).

We now have an interesting interpretation of this method: the density estimation by the kernel method can be obtained by smoothing observations with a P-coding originating with the kernel K.

Example

Say $n=3$ and $x_1=4$, $x_2=12$, $x_3=10$ are three observations of X. The empirical distribution function is the discontinuous function:

$$F_3(\omega; x) = 1/3 \sum_k 1_{]-\infty,\, x]}(x_k)$$

If we code these observations using the rectangular kernel $K = 1_{[-1/2;\, 1/2]}$ with for example $h(3)=3$ we obtain the continuous function:

$$G_3(\omega; x) = 1/3 \sum_k \phi^3_{]-\infty,\, x]}(x_k)$$

with

$$\phi^3_{]-\infty,\, x]}(y) = \begin{cases} 0 & \text{if } y < x - 3/2 \\ 1/3(y-x) + 1/2 & \text{if } x - 3/2 \leq y \leq x + 3/2 \\ 1 & \text{if } y > x + 3/2 \end{cases}$$

See Figure 5.2, where F_3 and G_3 are plotted.

6. THE APPLICATION OF P-CODING TO THE ESTIMATION OF PCA OF A p-DIMENSIONAL r.v.

Let $X = (X_1, \ldots, X_p)$ be a centred p-dimensional r.r.v. and let $(X^1, \ldots, X^n)$ be a sample of X, observed in ω.

This defines a datamatrix $Z = {}^t(X^1(\omega), \ldots, X^n(\omega)) = (x_{ki})$ with $x_{ki} = X_i^k(\omega)$, which can be displayed as a cloud of n points in the space R. We note x_k, the p-vector $(x_{k1}, \ldots, x_{kp})$. The PCA of this cloud of points consists of seeking the linear combination of the orthogonal r.v. X_i, which has a maximal variance. We know that PCA is obtained by diagonalizing the variance–covariance matrix $V = {}^tZDZ$, where D is the matrix of the metric chosen (cf. Chapter 6).

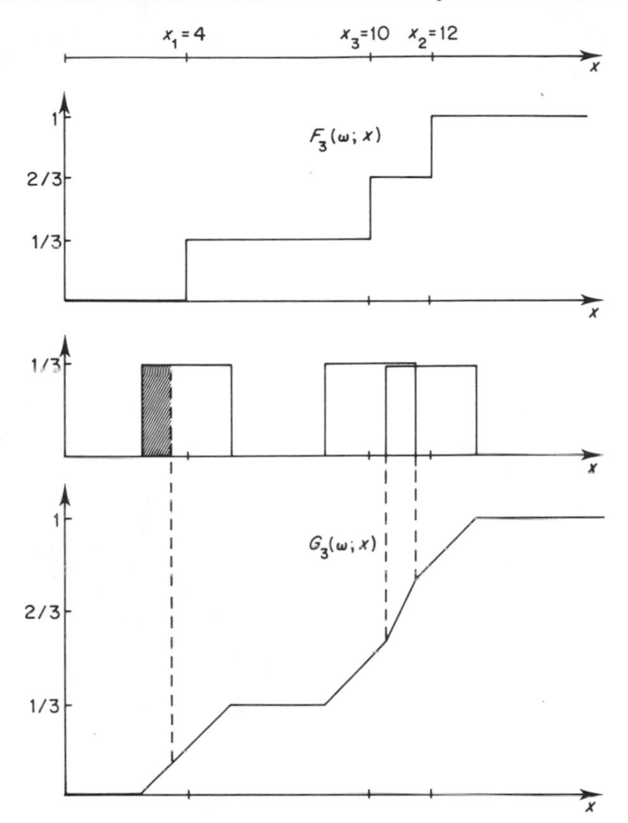

Figure 5.2. Continuous (G_3) and discontinuous (F_3) estimation of the distribution

We assume from now on that $D=n^{-1}I$, thus $V=n^{-1t}ZZ$. We can see that the PCA of the data matrix is actually the PCA of the distribution: $\delta(\omega)=n^{-1}\sum_k \delta_{xk}$, the distribution on the space R^P; $B_R P$. Indeed, the PCA of this distribution is obtained by diagonalizing its variance–covariance matrix, with the leading term

$$v_{ij}=\int_R u_i u_j \, d\delta(u_1, \ldots, u_p)=n^{-1}\sum_k x_{ki}x_{kj}.$$

We observe that the PCA of the data matrix of data is also the PCA of the empirical mean of the sample coded by the Dirac coding: $(\delta_{x^1(\omega)}, \ldots, \delta_{X^n\omega})=(\delta_{x1.}, \ldots, \delta_{xn.})$. When, for some reason, we code the data with a P-coding P_X, we have in the space R^P a cloud of n p-dimensional distributions $P_{X^1(\omega)}, \ldots, P_{X^n(\omega)})$. The PCA of this cloud is then the PCA of the empirical mean of the coded sample:

$$v(\omega)=n^{-1}\sum_k P_{X^k(\omega)}=n^{-1}\sum_k P_{xk.}.$$

This PCA is obtained by diagonalizing the variance–covariance matrix of $v(\omega)$ with the leading term

$$w_{ij}=n^{-1}\sum_k \int u_i u_j \, \mathrm{d}P_{(Xkl,\ldots,Xkp)}(u_1,\ldots,u_p).$$

Consider the example shown in Figure 5.3 with $p=2$ and $n=8$.

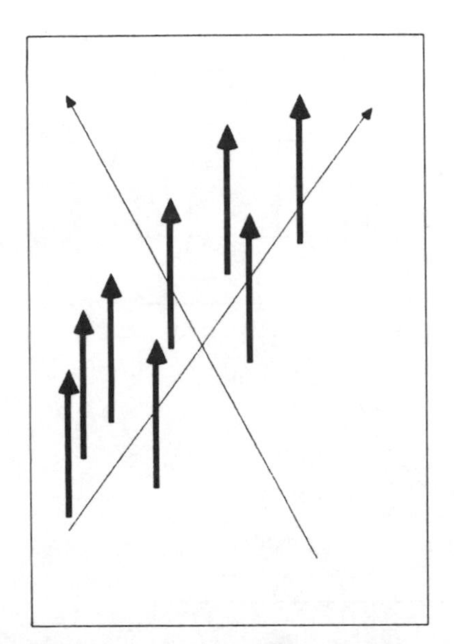

Figure 5.3. PCA with a density smoothing. (*Left*) PCA of the uncoded data $(x_{1.},\ldots,x_{8.})$ equivalent to the PCA of the distribution $\delta=1/8\sum_i \delta_{xi.}$. (*Right*) PCA of the coded data. $(\delta_{x1.}*\pi,\ldots,\delta_{x8.}*\pi)$ equivalent to the PCA of the distribution $v=1/8\sum_i \delta_{xi.}*\pi$.

Let us further this approach using the P-coding $P_x=\pi*\delta_x$ (π is a centred distribution on R^P). The leading term of the variance–covariance matrix of P_x is:

$$\alpha_{ij}=\int_P u_i u_j \, \mathrm{d}(\pi*\delta_x)=\int_{v+w\in R^P}(v_i+w_i)(v_j+w_j)\,\mathrm{d}\pi(v)\,\mathrm{d}\delta_x(w)$$

$$\int_{v\in R^P}(v_i+x_i)(v_j+x_j)\,\mathrm{d}\pi(v)=x_i x_j+\int_{R^P} v_i v_j \,\mathrm{d}\pi(v)$$

(because $\pi*\delta_x$ is centred, $\int_{R^P} v_i \,\mathrm{d}\pi(v)=0$).

Then, if W is the variance–covariance of the distribution π, the variance–covariance of $v(\omega)$ equals $n^{-1}\sum_k \pi*\delta_{Xk.}=V+W$.

Example 1

We can use the P-coding just as in Section 5, in order to estimate the density $f_x(u_1, \ldots, u_p)=f_x(u)$ of the distribution of X. The P-coding, which depends on the sample size, is $P_x=\pi_n * \delta_x$. Suppose that W_n, variance–covarance matrix of π_n, converges to the null matrix (variance–covariance matrix of δ_0). Then the PCA of $v(\omega)$ converges, when n tends to infinity, to the PCA of $\delta(\omega)$, which proves that the PCA of the coded sample is really an estimation from a sample of the PCA of X.

Example 2

Suppose that each observation $x_{k.}=(x_{k1}, \ldots, x_{kp})$ is tainted with error. And suppose also, in order to simplify, that the errors on each component $x_{k1}, \ldots, x_{kp}$ are uncorrelated. We can smooth, to give an account of this error, each component x_{kj} by a code $P_{xkj}=\pi_j * \delta_{xkj}$. Let π the distribution $\pi_1 \times \pi_2 \times \ldots \times \pi_p$ and σ^2 the variance of π_j. Then the P-coding is $P_x=\pi * \delta_x$ and its variance–covariance matrix is $V+W$, W being the diagonal matrix diag (σ^2). The PCA of the coded data is hence obtained by diagonalizing the matrix $V+W$.

If all the variances σ^2 are equal (each component having the same precision), we can observe that $V+W=V+\sigma^2 I$, while I is equal to the identity matrix. Then, the eigenvectors of $V+W$ are the same as those of V: the 'primary' data $(x_{1.}, \ldots, x_{n.})$ and the coded data $(P_{x1.}, \ldots, P_{xn.})$ have identical principal components.

We must notice that the eigenvalues are, however, not the same: when we give an account of the errors, the variances explained by the principal components are modified. To be more precise, if $\lambda_1, \lambda_2, \ldots, \lambda_p$ $(\lambda_1 \geq \lambda_2 \ldots \geq \lambda_p \geq 0)$ are the eigenvalues of V then the eigenvalues of $V+W$ are $\lambda_1+\sigma^2, \ldots, \lambda_p+\sigma^2$.

Let us compare the percentage of variance explained by the first j principal components of both the coded and not-coded data:
For the not-coded data, this percentage is:

$$\alpha_j=\frac{\sum_{k=1}^{j} \lambda_k}{\sum_{k=1}^{p} \lambda_k}$$

and for the coded data:

$$\beta_j=\frac{\sum_{k=1}^{j} \lambda_k+j\sigma^2}{\sum_{k=1}^{p} \lambda_k+p\sigma^2}$$

We can see that $\alpha_j \geq \beta_j$ indeed:

$$\alpha_j-\beta_j=\sigma^2 D^{-1}[p(\lambda_1+\cdots+\lambda_j)-j(\lambda_1+\cdots+\lambda_p)] \text{ (with } D=\sum_k \lambda_k)(\sum_k \lambda_k+p\sigma^2))$$

has got the same sign as:

$$(p-j)(\lambda_1+\cdots+\lambda_j)-j(\lambda_{j+1}+\cdots+\lambda_p)$$

Now, $\lambda_1 \geq \cdots \geq \lambda_j \geq \lambda_{j+1} \ldots \geq \lambda_p$. Then $(p-j)(\lambda_1+\cdots+\lambda_j) \geq (p-j)j\lambda_j \geq j$ $(\lambda_{j+1}+\cdots\lambda_p)$, and $\alpha_j-\beta_j$ is positive.

We can conclude, in this particular case, that if we code the data, the principal components and the representation of the cloud of points do not change, but the quality of the representation with the first j principal components in terms of explained variance is less good.

Let, for example, $\lambda_1=4$, $\lambda_2=3$, $\lambda_3=2$, $\lambda_4=1$ and $\sigma^2=1$:

J	1	2	3	4	
α_j	40	70	90	100	(in percentage)
β_j	36	64	86	100	

Component and Correspondence Analysis
Edited by J.L.A. van Rijckevorsel and J. de Leeuw

Chapter 6

Two Techniques: Monotone Spline Transformations for Dimension Reduction in PCA and Easy-to-Generate Metrics for PCA of Sampled Functions

Suzanne Winsberg
IRCAM, Paris, France

1. INTRODUCTION

The two techniques discussed in this chapter are presented together primarily because the author contributed to the development of both of them. They will therefore be presented independently and sequentially. Each of these techniques is related to different parts of this book. Just how each technique is related to material presented in other sections will be spelled out as part of the presentation. The first part of this chapter is devoted to monotone spline transformations for dimension reduction and the second part of the chapter treats special metrics designed for the analysis of sampled functions.

2. MONOTONE SPLINE TRANSFORMATIONS FOR DIMENSION REDUCTION

In the first of the two subjects presented in this chapter the data under consideration arise from the measurements of n objects with respect to p variables. This type of data may of course be displayed in an $n \times p$ matrix. Bilinear or factor analytic methods are often used to reveal relationships among the objects and among the variables by representing them in a space of low dimensionality. This point is to retain enough information for adequate representation of the original data while gaining simplicity for visualization and interpretation. The dimension reduction techniques presented in this chapter assume that non-linear relationships among the observed variables are required to adequately express the redundancy in the original data. In classical approaches to bilinear methods (Eckart and Young, 1936; Harman, 1967; Thurstone, 1947) reduction is based on the strong assumption that the redundancy in the original data may be expressed in terms of only linear relationships. Elsewhere in this book non-linear transformations of each column of the data matrix are considered to either achieve a greater reduction in the number of dimensions used to represent the objects or to achieve a more adequate representation of the original data than would be possible by requiring the transformations to be linear. However, sometimes the objects are measured on an ordinal scale. Thus it may be of interest to consider an approach for recovering empirical monotone transformations of the data giving rise to a minimum dimensional representation. Moreover non-monotone transformations have the disadvantage that in NCA (non-metric components analysis) the space is transformed; and in order to understand the resultant fit, we would need to transform back to the original space. This can only be achieved if the transformations are restricted to monotone functions. In this chapter we shall therefore restrict our attention to monotone transformation only.

Therefore it will be assumed that a monotone transformation of the values in each column of the data matrix is determined by a linear combination of a new set of values that characterize each row object. The monotone transformation is represented by a non-negative linear combination of integrals of basis splines. Monotone splines were first used by Winsberg and Ramsay (1980; 1981) to represent monotone transformation in regression analysis and in the analysis of pairwise preference data. Monotone splines were then used by Winsberg and Ramsay (1982; 1983) in dimension reduction techniques.

2.1. Two approaches to the dimension reduction problem

Two approaches to the dimension reduction problem are presented. The first is an extension of the fixed factor scores model of the Eckart–Young approximation (Eckart and Young, 1936). Let $\mathbf{X}$ denote the $n \times p$ matrix of original observations and $\mathbf{Y}(\mathbf{X})$ denote the matrix of transformed observations with entries $y_j(x_{ij})$ where the function $y_j(\cdot)$ is monotone. This model, denoted the $EY-T$ model proposes that the ith row of $\mathbf{Y}(\mathbf{X})$ is an observation from a multinormal distribution with mean $\boldsymbol{\mu}_i$ and variance–covariance matrix $\sigma^2\mathbf{I}$. The mean vector is further specified by the model $\boldsymbol{\mu}_i = \mathbf{A}\mathbf{f}_i$, $\mathbf{f}_i$ being an r-dimensional vector of factor scores and $\mathbf{A}$ being a p by r factor pattern matrix with $r \ll p$. Thus $\mathbf{Y}(\mathbf{X})$ is to be fitted by $\hat{\mathbf{Y}} = \mathbf{F}\mathbf{A}^t$ where the ith row of $\mathbf{F}$ is $\mathbf{f}^t_i$. $\mathbf{F}$ and $\mathbf{A}$ are determined only to within a non-singular linear transformation and its inverse respectively. Authors other than Winsberg and Ramsay (1983) who have considered the $EY-T$ model are Kruskal and

Shepard (1974) who discussed the fitting of monotone regression transformations; Young, Takane and de Leeuw (1978) and Gifi (1981a) who extended this approach to unordered data, and van Rijckevorsel (1982, 1987) who considered linear combinations of B-splines for $y_j(\cdot)$. The approach discussed in this chapter is the only one which considers smooth differentiable functions guaranteed to be monotonic. The last feature is the main departure from other work described in this volume. In addition in this chapter maximum likelihood estimation will be presented.

The second approach to the dimension reduction approach presented in this chapter corresponds to the random factor scores model denoted the $MN-T$ model. Here the assumption is that each row of $\mathbf{Y}(\mathbf{X})$ is a randomly sampled observation from a multinormal distribution with a variance–covariance matrix Σ, so that the transformation $y_j(\cdot)$ are designed to render the data multinormally distributed. This approach using monotone splines was exploited by Winsberg and Ramsay (1983). An earlier paper by Andrews, Gnanadesikan and Warner (1971) considered power transformations to bivariate normality.

2.2. I-splines

This section of the chapter is devoted to a brief overview of monotone splines denoted as I-splines. Since B-splines are discussed earlier in this volume and M-splines are proportional to B-splines ($M_{qk}=\Delta_{qk}B_{qk}$, where $\Delta_{qk}=(t_{q+k}-t_q)/k$) only monotone or I-splines will be discussed in this chapter. Monotone splines were first used and defined by Winsberg and Ramsay (1980, 1981). A complete discussion of their properties may be found in Winsberg and Ramsay (1983).

Since both the M-spline and B-spline bases are everywhere positive or zero, in applications requiring the generation and fitting of monotone spline functions it is natural to consider as a basis integrated M- or B-splines, in particular the family of I-splines defined as

$$I_{qk}(x)=\int_a^x M_{qk}(u)\,\mathrm{d}u \tag{2.1}$$

where the splines are defined on an interval $[a, b]$ and k is the order of the splines. Because of the relation between the M-spline and its derivative it can be shown that I-splines may be specified by

$$I_{qk}=\begin{cases} 0, & q>s \\ \sum_{i=q}^{s} \Delta_{i,k+1}M_{i,\,k+1}(x), & s-k+1\leq q\leq s \\ 1, & q<s-k+1 \end{cases} \tag{2.2}$$

where s is defined by $t_s\leq x<t_{s+1}$. Thus an I-spline of order k is a piecewise polynomial of degree k. In the useful special case of $k=2$

$$I_{s-1,\,2}(x) = 1 - (\Delta_{s1}\Delta_{s-1,2})^{-1}(t_{s+1} - x)^2/2$$
$$I_{s,\,2}(x) = (\Delta_{s1}\Delta_{s2})^{-1}(x - t_s)^2/2. \tag{2.3}$$

Although I-splines do not have a banded coefficient matrix they are bounded since $\int_a^b M_{qk}(u)\,du = 1$ and they are constant everywhere except within a limited interval.

I-splines are most interesting from a statistical point of view since they have the characteristics of a probability distribution function. The distribution defined by the qth I-spline has mean and variance

$$E_q(x) = \int_a^b x\,dI_q = \bar{t}_q = (t_q + \cdots + t_{q+k})/(k+1)$$
$$\text{Var}_q(x) = [(k+1)^2(k+2)]^{-1} \sum_{j=q}^{q+k} \sum_{i>j}^{q+k} (t_i - t_j)^2. \tag{2.4}$$

Therefore $F(x) = \sum_q c_q I_{qk}(x)$ is a mixture of distributions when $c_q \geq 0$ and $\sum c_q = 1$.

To conclude this section some general remarks on the choice of order and knots are in order. The total flexibility of the fitted curve depends primarily on the dimensionality of the space of splines m. When there is one knot per junction point this is the sum of the order k and the number of interior knots $L-1$. When there are no interior junction points and $m = k$ a polynomial of degree k is fit to the data. When $k = 0$ the fitting function is a histogram. In the extreme case where there is a junction point at each data point one obtains monotonic regression. That is for $k = 0$ the M-spline is zero almost everywhere except at the knots, which makes the I-spline an increasing step function. The problem with polynomials is that the value of any parameter affects the shape of the polynomial everywhere and conversely its estimate depends upon all the data. On the other hand it is highly continuous. The problem with a histogram, on the other hand, is its discontinuity which makes it unappealing as a model for the data. Splines obviously permit any degree of compromise between continuity on the one hand and local influence of parameters on the other. The parameter c_q in $g(x) = \sum c_q M_{qk}(x)$ determines the contribution to the fitting function by the qth M-spline whose possible influence is limited to k intervals. The smaller k, the more local its influence and consequently the smaller the potential covariance between estimates of a pair of coefficients. From a statistical point of view this is generally desirable and would argue for the smallest possible value of k consistent with whatever continuity requirements hold.

The desirability of small order is offset somewhat by the problem of dependence on knot placement. While the fitting problem is linear in the coefficients c_q, it is highly non-linear in the knots t_q, and therefore it is desirable to avoid much optimization with respect to them. When one knows in advance that much of the curvilinearity will fall within a certain region, this is not a problem. It is also not a serious problem when one has enough data to place a liberal number of knots in any region. Winsberg and Ramsay (1980, 1981) have found that in the presence of mild curvilinearity equal spacing of knots and $k=2$ works very well and that, even when curvilinearity is more local, one can do as well by readjusting the knots a few times with crude procedures and keeping $k=2$ as one can using higher orders. It must also be added that a reasonable number of data should be in the vicinity of any interior knot.

2.3. Choice of fitting criterion

Since one of the departures in this chapter from techniques presented elsewhere in this book is the use of maximum likelihood in addition to least squares estimation, it seems appropriate at this point to make some remarks comparing these two criteria. When estimated transformations are to be applied to one or more dependent variables, the question of choice of fitting criterion arises. In this section maximum likelihood and least squares are first compared from a classical sampling theory point of view. Within this framework it is natural to talk of a 'true' or population transformation which is being estimated and to compare the two types of estimate. It is also possible to compare the two on a non-probabilistic basis and this section concludes with this type of comparison.

Both observations x_{ij}, $i=1, \ldots, n$; $j=1, \ldots, p$, and their transformed values $y_j(x_{ij})$ have some distribution conditional on the values of a model function $\hat{y}_{ij}(\boldsymbol{\theta})$, where $\boldsymbol{\theta}$ is the structural parameter vector to be estimated. The data transformation $y_j(\cdot)$ is determined by an estimated parameter vector $\mathbf{c}_j$, and can also be written as $y(x; \mathbf{c}_j)$ to make its dependence on $\mathbf{c}_j$ explicit. The two classical approaches to describing the random behaviour in the data are (a) to assume that the $y_j(x_{ij})$'s have independent identical distributions about the $\hat{y}_{ij}$'s, and (b) to assume that the x_{ij}'s have independent identical distributions about the $y_j^{-1}(\hat{y}_{ij})$'s. The two approaches will not be equivalent in general and some thought must be directed towards the question of which is the more natural. Process (a), the 'model-error-transformation' formulation, is attractive where the dependent variables have been selected from a number of potential variables non-linearly but monotonically related to each other. Such is very often the case in the social sciences, where dependent variables are constructed from responses such as category choices to which numbers have been more or less arbitrarily assigned. Then any monotonic transformation is about as meaningful as any other, except possibly for smoothness considerations. The approaches of Box and Cox (1964) and Winsberg and Ramsay (1980, 1982, 1983) imply this formulation. Process (b), the 'model-transformation-error' formulation, would be particularly appropriate where the dependent variable is uniquely significant in its original form, and the results become difficult to interpret if its metric is altered by a non-linear monotone transformation.

If it is assumed according to the 'model-error-transformation' approach that the $y_j(x_{ij})$'s are randomly distributed about the values of $\hat{y}_{ij}(\boldsymbol{\theta})$, it is none the less necessary to rephrase this formulation so as to explicate the behaviour of that which is actually observed, namely the x_{ij}'s. To this end, let the density function of $y_j(x_{ij})$ be $f[y_j(x_{ij}) \mid \hat{y}_{ij}(\boldsymbol{\theta}), \sigma]$, where σ is a dispersion parameter. Then the density function for x_{ij} is given by multiplying f by the jacobian of the transformation to yield $f[y_j(x_{ij}) \mid \hat{y}_{ij}(\boldsymbol{\theta}), \sigma]D_x y_j(x_{ij})$, assuming that transformation $y_j(\cdot)$ is differentiable and monotone increasing. Given this density function for the x_{ij}'s themselves, the likelihood function can then be computed in the usual way. Conversely, one may begin by supposing that the x_{ij}'s have a density function $g[y_j(x_{ij}) \mid \hat{y}_{ij}(\boldsymbol{\theta}), \sigma]$ from which it follows that the transformed variables will have density $[D_x y_j(x_{ij})]^{-1} g[y_j(x_{ij}) \mid \hat{y}_{ij}(\boldsymbol{\theta}), \sigma]$. Note, however, that this introduces a dependency of the density on the shape of the transformation, and it may be difficult to give this any intuitive meaning in many applications.

In order to express these alternative approaches more concretely, let us suppose that the $y_j(x_{ij})$'s are normally distributed over replications with expectation $\hat{y}_{ij}(\boldsymbol{\theta})$ and standard deviation σ. If the transformations $y_j(\cdot)$ are in the class of monotone I-spline functions of order k discussed in the previous section, then the jacobian $D_x y_j(\cdot)$ is a non-negative linear combination of M-splines and hence everywhere non-negative. The log likelihood is then given by

$$
\begin{aligned}
\log L &= \sum_i \sum_j \{\log f[y_j(x_{ij}) \mid \hat{y}_{ij}(\boldsymbol{\theta}), \sigma] + \log D_x y_j(x_{ij})\} \\
&= -np \log \sigma - (2\sigma^2)^{-1} \sum_i \sum_j [y_j(x_{ij}) - \hat{y}_{ij}(\boldsymbol{\theta})]^2 \qquad (2.5) \\
&\quad + \sum_i \sum_j \log \sum_q c_{jq} M_{qk}(x_{ij}).
\end{aligned}
$$

Note that the log likelihood is comprised of the familiar error sum of squares term plus a term determined by the shape of the transformations. It is this final term that provides the essential different between maximum likelihood estimation and least squares estimation.

The 'model-transformation-error' formulation, on the other hand, makes the observed x_{ij}'s the dependent variables rather than any estimated transformations of them. Consequently the addition of a jacobian term in the log likelihood is not required. The assumption that x_{ij} is independently and identically normally distributed over replication about $y_j^{-1}[\hat{y}_{ij}(\boldsymbol{\theta})]$ then results in maximum likelihood estimation of $\boldsymbol{\theta}$ and the $\mathbf{c}_j$'s being equivalent to least squares fitting.

From a computational point of view the two approaches have somewhat different characteristics. The 'model-error-transformation' approach is usually non-quadratic in the $\mathbf{c}_j$'s due to the jacobian term, but may be comparatively simple to solve in terms of $\boldsymbol{\theta}$. On the other hand, the 'model-transformation-error' formulation can be rather simple in terms of the $\mathbf{c}_j$'s but more complex in terms of solving for $\boldsymbol{\theta}$. Thus, when vector $\boldsymbol{\theta}$ is large relative to the size of $\mathbf{c}_j$, an alternating maximum likelihood algorithm may be more attractive for the former than for the latter approach.

Even if one does not wish to motivate the choice of fitting criterion by probabilistic considerations, one is still required to make a choice. Just as the normal distribution may be completely inappropriate in some applications, so, too, may the least squares criterion be a poor choice. Consider the following example. Suppose one has a set of observations consisting of the pairs (u_i, v_i), $i=1, \ldots, n$, $v_i>0$, and it is desired to fit the model $au+b$ to the values v^c, where a, b and c are to be estimated. Then $a=0$, $b=1$ and $c=0$ is a solution to the least squares problem for any set of observations. However, the loss function

$$Q=\sum (v_i^c=au_i+b)^2-\sum \log(cv_i^{c-1})$$

does not admit this trivial solution. The difficulty with least squares arises because both the model and the fitted values can be compressed to the same value by a particular choice of parameters. This tendency of least squares fitting to result in compression of the dependent variables has been documented by Ramsay (1977). Alternatively the criterion

$$Q'=\sum [v_i-(au_i+b)^d]^2$$

avoids this compression problem Note that $Q''=\sum [v_i^c-(au_i+b)]^2$ is particularly inappropriate. In general the choice between these criteria depend upon the problem under consideration. The choice of Q'' may be suitable under certain circumstances.

2.4. Parameter estimation

The parameters to be estimated in the $EY-T$ and $MN-T$ models fall into two groups: the transformation parameter vectors $\mathbf{c}_j$ and the structural parameters consisting of the factor pattern matrix $\mathbf{A}$ and, in the case of the $EY-T$ model, the factor scores $\mathbf{F}$. The $EY-T$ model also requires a scalar σ determining the dispersion of the residuals. The transformation for variable j is

$$y_j(x)=c_{j0}+\sum_{q}^{m_j} c_{jq}I_{qk}(x), \qquad c_{jq}\geq 0 \text{ for } q\geq 1 \tag{2.6}$$

where I-spline I_{qk} was defined in (2.1) and is determined by a knot sequence $\mathbf{t}_j$ within an interval $[a, b]$ and by an order k. The derivative of the transformation is $D_x y_j(x)=\sum c_{jq}M_{qk}(x)$.

There are two types of indeterminacy in the parameters that must be resolved. The origin of each transformed variable is indeterminate, and thus the constants c_{j0} can be chosen so that each column of $\mathbf{Y}(\mathbf{X})$ has a mean of zero. For the $EY-T$ model this implies in addition that the factor scores matrix $\mathbf{F}$ is also column-centred. The scale of each transformed variable is also indeterminate since any change of scale for $y_j(\cdot)$ can be compensated for by a change in scale in the jth column of $\mathbf{A}$. Consequently, the transformed variables can be required to satisfy the restrictions $\mathbf{1}'\mathbf{c}_j=1, j=1, \ldots, p$ where $\mathbf{c}_j=\{c_{j1}, \ldots, c_{jm}\}^t$. Other means of resolving the location and scale indeterminacies are possible and may be suitable for certain applications.

The algorithm employed for both models requires an alternation between maximizing the log likelihood with respect to the transformation parameters and with respect to the structural parameters. Although this procedure can converge linearly at best, it is simple to program, requires a modest amount of computer core memory, and has converged at an acceptable rate in Winsberg and Ramsay's experience. The solution for the structural parameters given the transformation parameters is explicit and therefore only the transformation estimation phase requires numerical optimization technology. The log likelihood are given here for each model. For a complete description of the algorithm see Winsberg and Ramsay (1983). Let matrices $\mathbf{I}_j$ and $\mathbf{M}_j$ be the n by m_j matrices containing the values of the I-splines and M-splines, respectively, for each observation on the jth variable. Then the n values of the transformed j^{th} variable are contained in $\mathbf{I}_j\mathbf{c}_j$ within a constant term and the values of its derivative in $\mathbf{M}_j\mathbf{c}_j$. It will be assumed that the columns of $\mathbf{Y}(\mathbf{X})$ have been centred, and this matrix will be referred to for simplicity as $\mathbf{Y}$.

$EY-T$ model: In this model it is assumed that each $y_j(x_{ij})$ is a sample of size one drawn from a normally distributed population with mean $\hat{y}_{ij}$ and standard deviation σ. The n by p matrix of mean values $\hat{\mathbf{Y}}$ is specified by the model $\hat{\mathbf{Y}}=\mathbf{F}\mathbf{A}^t$. As was described in Section 2.3, this implies that the log likelihood is

$$\log L = -np\log\sigma - (2\sigma^2)^{-1}\text{tr}[(\mathbf{Y}-\hat{\mathbf{Y}})^t(\mathbf{Y}-\hat{\mathbf{Y}})] + \sum_j \mathbf{1}_n^t \log(\mathbf{M}_j\mathbf{c}_j). \quad (2.7)$$

$\mathbf{F}$ and $\mathbf{A}$ are then given by the usual Eckart–Young decomposition process in which the singular value decomposition $\mathbf{Y}=\mathbf{U}\mathbf{D}\mathbf{V}^t$ is computed and $\mathbf{F}$ is taken to be proportional to those columns of $\mathbf{U}$ corresponding to the r largest singular values and $\mathbf{A}=\mathbf{Y}^t\mathbf{F}(\mathbf{F}^t\mathbf{F})^{-1}$. The estimate of σ^2 is $(np)^{-1}\text{tr}[(\mathbf{Y}-\hat{\mathbf{Y}})^t(\mathbf{Y}-\hat{\mathbf{Y}})]$.

$MN-T$ model: The assumption here is that the values of the transformed variables for the ith case are randomly sampled from a multinormal distribution with some mean vector $\boldsymbol{\mu}$ and variance–covariance matrix $\boldsymbol{\Sigma}$. If one denotes the sample variance–covariance matrix $n^{-1}\mathbf{Y}'\mathbf{Y}$ by $\mathbf{C}$, then the log likelihood is given by

$$\log L = -\left(\frac{n}{2}\right)(\log|\Sigma| + \text{tr}\,\boldsymbol{\Sigma}^{-1}\,\mathbf{C}) + \sum_j \mathbf{1}_n^t \log(\mathbf{M}_j\mathbf{c}_j) \quad (2.8)$$

where $\boldsymbol{\Sigma}_j^{-1}$ is the jth column of $\boldsymbol{\Sigma}^{-1}$ and σ^{jl} is the jl^{th} element of $\boldsymbol{\Sigma}^{-1}$, respectively. The variance–covariance matrix $\boldsymbol{\Sigma}$ may or may not be assumed to have some structure. If not, then its maximum likelihood estimate given the transformations is $\mathbf{C}$. However, dimension reduction can be achieved by assuming a factor structure. Given that the scale restrictions imposed on the transformations impose a sort of normalization on $\mathbf{Y}$, it may be plausible to consider the simple model $\boldsymbol{\Sigma}=\mathbf{A}\mathbf{A}'+\sigma^2\mathbf{I}$. If $\mathbf{C}$ has the decomposition $\mathbf{C}=\mathbf{V}\mathbf{C}\mathbf{V}^t$ then the single unique variance parameter σ^2 is estimated by

$$\hat{\sigma}^2 = (p-r)^{-1} \sum_{p-r+1}^{p} d_{jj}$$

and a factor pattern matrix element a_{jl} by $v_{jl}(d_u - \hat{\sigma}^2)^{1/2}$. More complex factor structures such as the usual unconstrained factor analytic model may also be proposed, but these will involve considerably more computational overhead.

The approach to the linearly constrained problem used by Winsberg and Ramsay was one in which at each iteration the parameters are linearly transformed and then a subset of them corresponding to active constraints are held fixed. An active constraint in this context occurs when a coefficient c_{jq} has become zero and the corresponding gradient element is negative. The equality constraint $\mathbf{1}^t_{cj} = 1$ must also be respected.

2.5. Examples

In each of the analyses presented here order two splines with one or two interior knots were used implying three or four transformation parameters per variable. In order to simplify the interpretation of knot values, it will be assumed that each untransformed variable $\mathbf{x}_j$ has been normalized so that its minimum value is zero and its maximum value is one, and that the lower and upper exterior knots are zero and one, respectively. Monotone transformations are particularly suitable for all of the examples presented in this chapter.

2.5.1. *Characteristics of* 39 *automobiles*

An example of a data set is the specifications on five variables for a sample of automobiles reviewed by *Consumer Report* in the April 1983 issue. The data are given in Table 6.1 and the interior knots used relative to a variation over [0, 1] are shown in Figure 6.2.

The rows of F for the two-dimensional $EY-T$ model are plotted in Figure 6.1 along with the direction representing the five variables. It is evident that the first factor can be labelled 'size', and that the second is primarily the differential between size and price, with the inexpensive economical cars having high scores on this factor. Figure 6.1 displays the estimated transformations resulting from the use of a single interior knot for each variable. The use of more knots did not alter these transformations significantly. We see that high prices, large engine sizes and large weights are relatively reduced by the transformations, but that city and expressway gas consumptions have more nearly linear transformations. Transformations estimated by the $MN-T$ model were very similar to these. Maximum likelihood estimation was used in each case.

2.5.2. *Characteristics of* 44 *automobiles*

The next illustration involves the analysis of the data in Table 6.2, again comprising measures on automobiles this time on 44 automobiles reported in

Table 6.1. *Consumer Report* specifications for 1983 automobiles

Automobile	*Symbol*	*Price* (100 $)	*Displacement* (*litres*)	*Gas consumption* *City* (*litres*/100)	*Expressway* (*km*)	*Weight* (100 *kg*)
Cavalier	A	62	1.8	12.3	5.3	11.1
Camaro	B	82	2.5	14.7	7.1	13.2
Chevette	C	56	1.6	12.3	6.9	9.7
Datsun 200SX	D	80	2.2	10.2	6.1	11.8
Honda Accord	E	86	1.8	10.7	5.8	10.0
Honda Civic	F	71	1.5	9.0	5.2	8.9
Ford Escort	G	69	1.6	10.2	5.2	9.4
Ford Mustang	H	76	2.3	15.6	6.7	12.0
Mazda GLC	I	67	1.5	11.2	5.2	9.1
Nissan Sentra	J	63	1.5	9.4	4.8	8.7
Nissan Stanza	K	72	2.0	10.2	5.2	9.8
Plymouth Horizon	L	63	1.7	9.0	5.4	9.9
Dodge Colt	M	57	1.4	10.2	5.1	8.8
Renault Alliance	N	72	1.4	11.2	5.2	9.3
Renault 18I	O	84	1.6	11.7	6.5	10.6
Renault Le Car	P	53	1.4	10.2	5.7	8.4
Subaru	Q	61	1.6	11.7	6.1	9.3
Toyota Corolla	R	65	1.8	11.2	6.0	10.2
Toyota Celica	S	87	2.4	11.2	5.8	12.0
Toyota Starlet	T	61	1.3	9.0	5.1	7.9
Toyota Tercel	U	61	1.5	10.2	5.1	9.7
Volks Jetta	V	82	1.7	11.7	5.8	9.9
Volks Rabbit	W	74	1.7	10.7	6.0	8.9
AMC Eagle	X	107	4.2	23.5	11.7	15.7
Buick Century	Y	102	3.0	15.6	7.3	13.0
Buick Regal	Z	105	3.8	18.1	7.8	15.0
Chev Celebrity	1	94	2.5	15.6	6.0	12.7
Chrysler E class	2	105	2.6	15.6	8.1	12.9
Datsun Maxima	3	114	2.4	15.6	8.4	13.3
Dodge Aries	4	86	2.2	13.0	6.9	11.1
Dodge Diplomat	5	94	3.7	23.5	10.2	15.5
Dodge 400	6	101	2.2	12.3	6.7	11.7
Ford Fairmont	7	79	2.3	16.8	8.4	13.1
Ford LTD	8	95	2.3	14.7	7.8	13.4
Pontiac Phoenix	9	86	2.5	14.7	6.9	12.1
Toyota Cressida	@	132	2.8	14.7	7.8	13.6
Chev Impala	#	101	5.0	19.6	7.5	16.9
Ford LTD Crown	$	111	5.0	19.6	9.0	17.6
Oldsmobile 98	%	133	4.1	19.6	8.7	18.3

Table 6.2. Automobile data from 1986 *Consumer Report*

Automobile	*Plot symbol*	*Price*	*Displ*	*City gas*	*Expressway gas*	*Wgt*	*Rel*	*Org*	*Type*
Chev Chevette	A	56	1.6	13.3	6.8	10.0	1	A	Sub
Chev Nova	B	74	1.6	10.5	5.5	10.2	5	AJ	Sub
Chev Spectrum	C	67	1.5	10.1	5.3	8.7	5	AJ	Sub
Dodge Colt	D	56	1.5	11.0	5.6	9.9	5	AJ	Sub
Dodge Omni	E	68	1.6	11.5	5.6	9.5	1	A	Sub
Ford Escort	F	61	1.9	12.0	6.2	10.9	3	A	Sub
Honda Civic	G	56	1.5	11.5	6.5	9.2	5	J	Sub
Mitsubishi Tredia	H	75	2.0	12.0	6.3	10.9	4	J	Sub
Nissan Sentra	I	56	1.6	10.5	5.6	9.5	4	J	Sub
Renault Alliance	J	60	1.4	12.0	5.6	9.1	1	AE	Sub
Subaru	K	80	1.8	12.0	5.6	10.4	5	J	Sub
Toyota Corolla	L	73	1.6	11.0	5.3	10.3	5	J	Sub
Toyota Tercel	M	56	1.5	11.0	5.5	9.7	5	J	Sub
Volkswagen Golf	N	74	1.8	12.0	6.3	10.1	3	E	Sub
Volkswagen Jetta	O	83	1.8	12.6	6.3	10.5	5	E	Sub
Chrysler Laser	P	94	2.2	15.8	7.4	12.7	1	A	Sport
Honda Civic CRX	Q	73	1.5	9.4	5.6	9.0	5	J	Sport
Honda Prelude	R	110	1.8	10.5	6.5	10.6	5	J	Sport
Isuzu Impulse	S	109	1.9	14.9	6.8	12.4	3	J	Sport
Mitsubishi Cordia	T	89	2.0	12.0	6.3	11.2	4	J	Sport
Nissan 200SX	U	95	1.8	12.6	6.5	12.2	4	J	Sport
Nissan Pulsar NX	V	87	1.6	9.7	5.1	9.2	4	J	Sport
Pontiac Fiero	W	89	2.5	12.6	6.7	11.5	1	A	Sport
Audi 4000S	X	142	2.2	14.1	8.2	10.7	4	E	Compact
Chev Cavalier	Y	67	2.0	15.8	7.2	11.6	1	A	Compact
Ford Tempo	Z	74	2.3	13.3	6.2	11.8	1	A	Compact
Honda Accord	1	88	2.0	13.3	6.3	11.8	5	J	Compact
Mazda 626	2	92	2.0	12.6	6.7	11.8	5	J	Compact
Mitsubishi Galant	3	132	2.4	14.9	6.7	12.9	3	J	Compact
Nissan Stanza	4	101	2.0	11.0	5.6	11.1	5	J	Compact
Olds Calais	5	93	2.5	14.1	6.7	12.0	3	A	Compact
Saab 900	6	126	2.0	14.1	7.9	12.9	3	E	Compact
Toyota Camry	7	97	2.0	11.0	5.5	12.2	5	J	Compact
Volvo	8	144	2.3	14.1	7.9	13.3	3	E	Compact
Buick Century	9	101	2.5	16.9	6.5	12.6	3	A	Medium
Chrysler Fifth Ave	0	149	5.2	23.0	9.7	16.2	1	A	Medium
Chrysler Le Baron	@	100	2.2	13.3	7.9	11.7	3	A	Medium
Dodge Aries	#	72	2.2	13.3	7.9	11.5	2	A	Medium
Dodge Lancer	$	94	2.2	15.8	8.7	12.7	3	A	Medium
Mercury Couger	%	114	3.8	15.8	8.7	14.1	3	A	Medium
Olds Cutlass Supreme	/	107	3.8	19.5	9.0	15.2	1	A	Medium
Pontiac 6000	&	95	2.8	16.9	8.7	12.6	2	A	Medium
Buick Electra	<	154	3.8	18.1	7.7	15.0	1	A	Full
Chev Caprice	>	106	5.0	18.1	8.2	16.5	2	A	Full

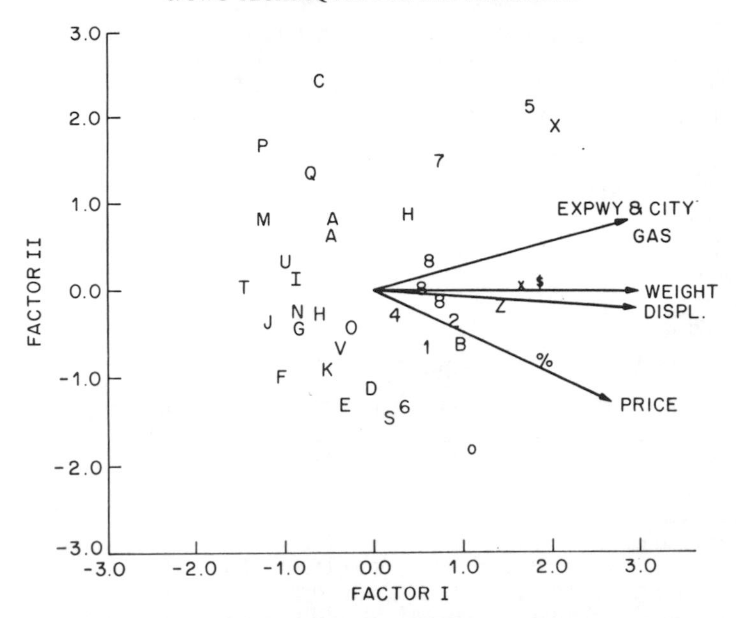

Figure 6.1. Biplot of the scores of the first two principal components of the transformed 1983 automobile variables. The $EY-T$ model in two dimensions with maximum likelihood estimation was used. Vectors indicate the directions corresponding to the transformed variables. The correspondence between symbols and automobiles is in Table 6.1

April 1986, issue of *Consumer Report*. This data was analysed by Ramsay using the following criterion:

$$\min_{C,F,A}\{t_r(Y-FA^t)^2\}$$

where

$$Y_{ij}=\sum_q c_{qj}I_{qj}(X_{ij}) \qquad i=1,\ldots,n, j=1,\ldots,p.$$

Figure 6.3 displays the five transformations resulting from the analysis in two dimensions. The strongly skewed variables, price and displacement, are transformed to have more symmetric distributions. Note also the bend in expressway gas not observable in the 1983 data. Figure 6.4 is a biplot of the first two principal components. Again the first clearly measures overall size while the second is primarily the differential between size and price. It might be described as the extent to which a car costs less and consumes more gas than its size would indicate. The results of the analysis are all in all quite similar to the results from the 1983 data despite the change in fitting criterion.

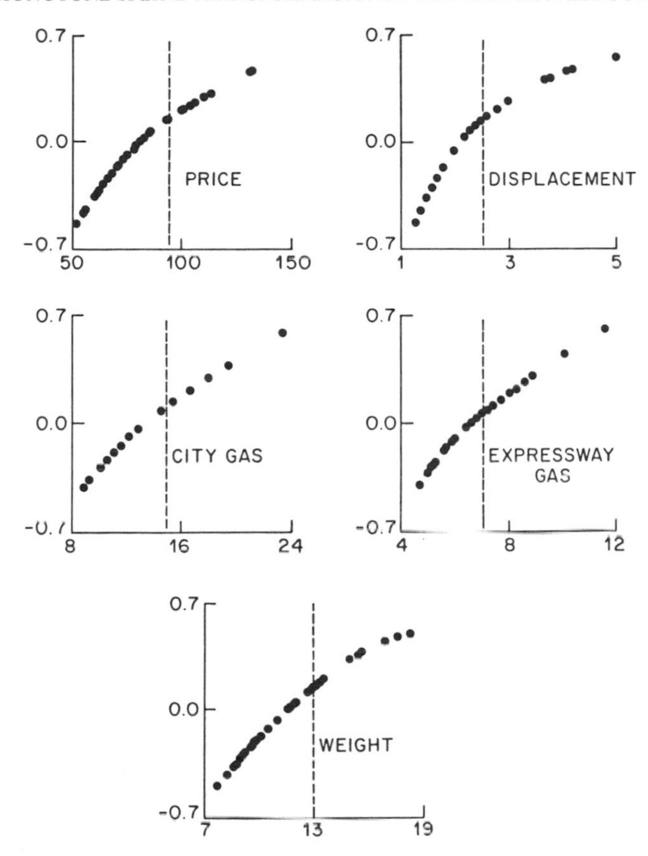

Figure 6.2. Estimated monotone spline transformations resulting from an approximation of the five transformed variables from the 1983 automobile data by the first two principal components ($EY-T$ model in two dimensions). The vertical dashed lines indicate the location of knots

2.5.3. *The spirits data*

The final example deals with the spirits data previously analysed using monotone spline transformations for a regression model in Winsberg and Ramsay (1980). The data were the log consumption of spirits, log real income, and log relative sprice of spirits in England for the years 1870 to 1938. The data were analysed using a two-dimensional $EY-T$ model with two interior knots and equal knot spacing for each variable. The signs of the second and third variables were reversed prior to analysis to produce monotone increasing transformations. The configuration of factor scores is shown in Figure 6.5 along with the vectors corresponding to the three transformed variables. It is evident that the horizontal direction represents consumption of spirits and the vertical direction the differential between income and price of spirits (or consumption). Thus there is a

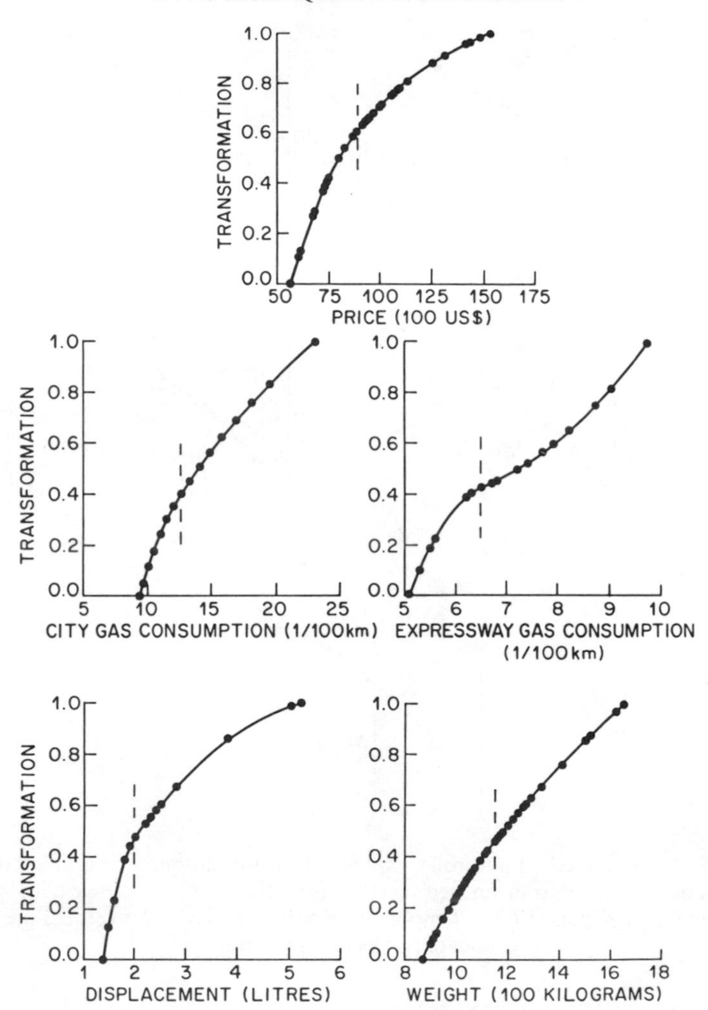

Figure 6.3. Biplot of the scores of the first two principal components of the transformed 1986 automobile variables. The $EY-T$ model in two dimensions with least squares estimation was used. Vectors indicate the directions corresponding to the transformed variables the correspondence between symbols and automobiles is in Table 6.2

steadily increasing gap between log real income and price in the years before the First World War and then an abrupt transition to the post-war period where the variation is primarily in terms of price. The corresponding transformations are presented in Figure 6.6 and resemble closely those obtained for the regression model, with the plateau in the transformation of log consumption corresponding to the initial years of the war being especially noteworthy. Analyses using other choices of knots including equal spacing yielded very similar results, as did the use of the $MN-T$ model.

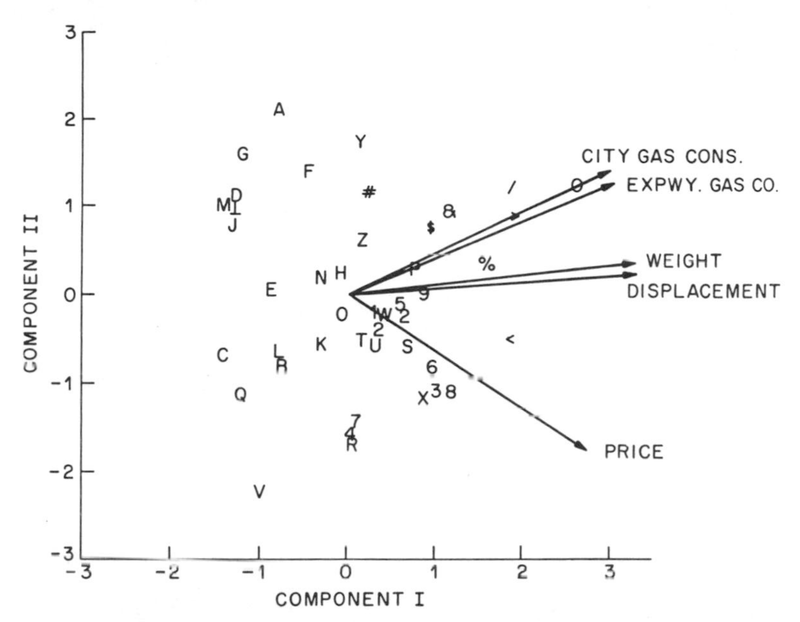

Figure 6.4. Estimated monotone spline transformation resulting from an approximation of the five transformed variables from the 1986 automobile data by the first two principal components ($EY-T$ model in two dimensions). The vertical dashed lines indicate the location of knots

2.6. Summary

Both the $EY-T$ and the $MN-T$ model are useful for gaining understanding of the structure of a data set. The $EY-T$ model is useful for displaying the relationship among the objects and among the variables in a space of greatly reduced dimensionality. The $MN-T$ model offers the advantage of its relative statistical generality. Finally monotone splines $I_q(x)$ are a most useful basis for representing monotone transformations. These results are easy to interpret. NCA analyses using B-splines can lead to non-monotone transformations which are difficult to interpret. (See the analysis of 1980 consumer reports car data by Rijckevorsel (1982, 1987), as compared with the results obtained by Winsberg and Ramsay, 1982.)

3. EASY-TO-GENERATE METRICS FOR PCA OF SAMPLED FUNCTIONS

In the second of the two subjects presented in this chapter the data arise from the sampling on n functions of time, $f_i(t)$, which may be periodic. Each function is sampled at p times, t_j, giving rise to a row-vector X_i of p values $x_i(t_j)=f_i(t_j)$. This vector $\mathbf{X}_i$ is called a sampled function. The data form an $n \times p$ matrix of values $X_{ij}(t_j)$.

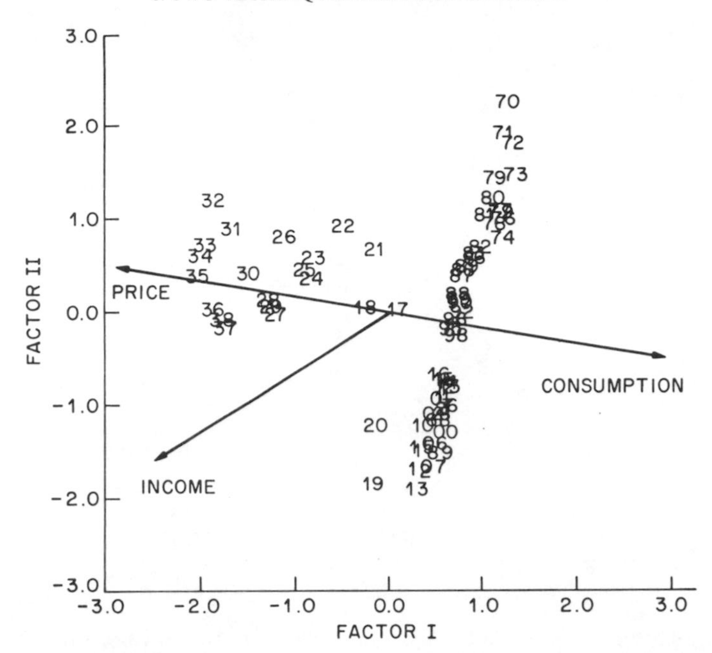

Figure 6.5. Biplot of the scores of the first two principal components of the transformed spirits variables. The $EY-T$ model in two dimensions with maximum likelihood estimation was used. Vectors indicate the directions corresponding to the transformed variables. The symbols correspond to the last two digits of the year of the observation

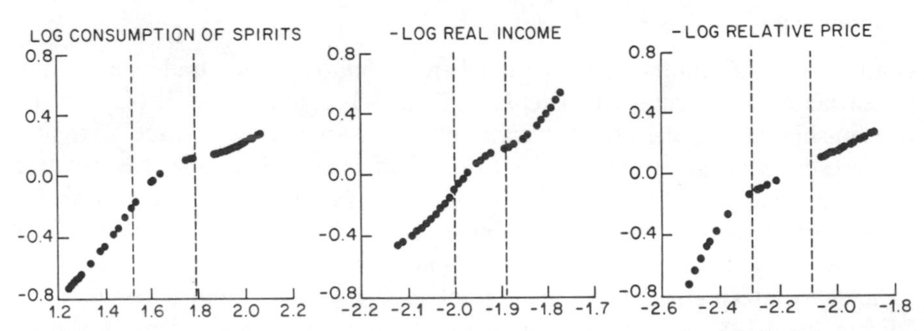

Figure 6.6. Estimated monotone spline transformations resulting from an approximation of the three transformed variables from the spirits data by the first two principal components ($EY-T$ model in two dimensions). The vertical dashed lines indicate the location of knots

The obvious or known variation in the data may often be described by a linear space of functions denoted S. Additional subtle variation in the data may be suspected and it is sometimes useful to remove the obvious or known variation in order to uncover the unknown or subtle variation. Consider the data set consisting of (periodic) mean monthly temperatures of thirty-two French cities shown in Table 6.3. The data look like sampled cosine functions with their peaks at July. The obvious variation is therefore the mean and a multiple of the cosine i.e., $S=\{1, \cos\}$. So one might want to remove this variation to study the residue, Res. The goal is then to filter out some known kinds of variation.

Consider methods to analyse X such as PCA, clustering or discriminant analysis. All of these methods are based on a quadratic metric between the row vectors of p dimensions. To attain the desired goal, a 'filter' is incorporated in the metric which removes the obvious or known variation.

Table 6.3. Mean monthly temperature (0.1°C) of French cities

City	*Plot symbol*	*Jan.*	*Feb.*	*Mar.*	*Apr.*	*May*	*June*	*July*	*Aug.*	*Sept.*	*Oct.*	*Nov.*	*Dec.*
Ajaccio	aj	77	87	105	126	159	198	220	222	203	163	118	87
Angoulème	ame	42	49	79	104	136	170	187	184	161	117	76	49
Angonne	an	46	54	89	113	145	172	195	194	169	125	81	53
Besançon	bes	11	22	64	97	136	169	187	183	155	104	57	20
Biarritz	bia	76	80	108	120	147	178	197	199	185	148	109	82
Bordeaux	bor	56	66	103	128	158	193	209	210	186	138	91	62
Brest	bre	61	58	78	92	116	144	156	160	147	120	90	70
Cler-Ferr.	clf	26	37	75	103	138	173	194	191	162	112	66	36
Dijon	dij	13	26	69	104	143	177	196	190	156	105	57	21
Embrun	emb	5	16	57	90	130	164	189	183	153	101	46	5
Grenoble	gre	15	32	77	106	145	178	201	195	167	114	65	23
Lille	lil	24	29	60	89	124	153	171	171	147	104	61	35
Limoges	lim	31	39	74	99	133	168	184	178	153	107	67	38
Lyon	lyo	21	33	77	109	149	185	207	201	169	114	67	31
Marseille	mar	55	66	100	130	168	208	233	228	199	150	102	69
Montpellier	mon	56	67	99	128	162	201	227	223	193	146	100	65
Nancy	ncy	8	16	55	92	133	165	183	177	147	94	52	18
Nantes	nan	50	53	84	108	139	172	188	186	164	122	82	55
Nice	nic	75	85	108	133	167	201	227	225	203	160	115	82
Nîmes	nim	57	68	101	130	166	208	236	229	197	146	98	65
Orléans	orl	27	36	69	98	134	166	184	181	156	109	66	36
Paris	par	34	41	76	107	143	175	191	187	160	114	71	43
Perpignan	per	75	84	113	139	171	211	238	233	205	159	115	86
Reims	rei	19	28	62	94	133	164	183	179	151	103	61	30
Rennes	ren	48	53	79	101	131	162	179	178	157	116	78	54
Rouen	rou	34	39	68	95	129	157	176	172	150	110	68	43
St Quen	stq	20	29	63	92	127	156	174	174	150	105	61	31
Strasbourg	str	4	15	56	98	140	172	190	183	151	95	49	13
Toulon	tou	86	91	112	134	166	202	226	224	205	165	126	97
Toulouse	tls	47	56	92	116	149	187	209	209	183	133	86	55
Tour	tou	35	44	77	106	139	174	191	187	162	117	72	43
Vichy	vic	24	34	71	99	136	171	193	188	160	110	66	34
Mean		39	48	81	109	143	177	198	195	169	123	79	48

3.1. Filtering techniques

As described elsewhere in this book, Besse and Ramsay (1986) studied 'filtering' for these techniques and developed a family of metrics which attain this goal. They dealt with PCA but their ideas apply equally well to other methods based on metrics. The Besse–Ramsay approach will be reviewed here in such a way and so as to contrast it with a simpler approach developed by Winsberg and Kruskal (1986). Conceptually Besse and Ramsay's approach is based on interpolation smoothing of the data, although in practice the interpolation/smoothing is not necessarily carried out.

When dealing with a set of sampled functions one can carry out ordinary PCA on X to obtain r functions and interpolate the resulting principal components. Besse (1979) suggested reversing these steps. That is, interpolate the rows of X to get continuous functions and carry out PCA on the functions to get r principal components. If the interpolation is carried out by splines, Besse (1979) developed a *metric M* such that a discrete PCA on X based on the metric M is equivalent to a PCA on the interpolated functions based on the identity metric. In practice one may obtain these results without reversing the steps by carrying out the PCA on X based on the metric M. One obtains r principal components which may then be interpolated. These arguments generalize to include smoothing without difficulty.

Besse and Ramsay's (the B–R) approach begins by choosing a differential operator that annihilates the functions to be removed. In the temperature example mentioned above the appropriate operator is $D+(1/\omega^2)D^3$ where D is d/dt. Now working with norms instead of metrics the B–R approach is to calculate and use the norm defined by these steps (i) interpolate or smooth the sampled functions; (ii) apply the operator; (iii) use the usual function space norm, L_2. In their method one must consider the reproducing kernel associated with the differential operator and calculate it.

Besse and Ramsay uses splines which minimize the L norm for interpolation. The function f_i for the points (t_j, X_{ij}) minimize $\int_0^T [f_i(t)]^2 dt$ subject to $f_i(t_j)=X_{ij}$. Note that the differential operator L that they use for interpolation is the same operator F that they chose as a filter. Unfortunately operators that are necessary as filters give rise to interpolated functions with undesirable properties. One should therefore choose L separately from F. ($L=D^2$ gives rise to cubic splines and this operator is desirable in many cases.) It is possible to generalize their method and find M so as to include the situation where L is different from F; but that becomes much more complicated. Moreover for periodic functions as in the case of the temperature data cited above, it is desirable that the interpolated function be periodic and possess the same order of continuity at the wrap around point as it has at the knots. This requirement also adds complications.

On the other hand filtering can be introduced without interpolation or smoothing. Moreover, in general interpolation alone changes the metric very little. Winsberg and Kruskal (1986) stimulated by Besse and Ramsay developed simpler approaches for filtering that do not use interpolation/smoothing. They work equally well and just as easily whether the data are periodic or not. The first approach developed by Winsberg and Kruskal is the discrete analog of the B–R approach and filters by the different operator that corresponds to the B–R differential operator. The second approach developed by Winsberg and Kruskal (W–K) filters by projection onto the space orthogonal to the space of functions to be removed.

The first W–K approach (the difference operator approach) may be illustrated as follows: to annihilate $S=\{1, \cos\}$ the differential operator is $D+(1/\omega^2)\,D^3$ which annihilates the larger space $\{\sin, \cos, 1\}$. In this case one uses the filter $F=\Delta+(1/\omega^2)\Delta^3$ where for periodic or non-periodic functions respectively when $p=4$.

$$\Delta=\begin{bmatrix} -1 & 0 & 0 & 1 \\ 1 & -1 & 0 & 0 \\ 0 & 1 & -1 & 0 \\ 0 & 0 & 1 & -1 \end{bmatrix} \qquad \Delta=\begin{bmatrix} -1 & 0 & 0 \\ 1 & -1 & 0 \\ 0 & 1 & 1 \\ 0 & 0 & 1 \end{bmatrix}$$

In the second W–K approach (projection operator approach) a basis for S is chosen. These functions need not be orthogonal. The matrix K is constructed such that the rows consist of the p values for the functions to be removed. These functions are called the null functions. The filter is then defined as $F=\mathrm{I}-K^0\equiv\mathrm{I}-K^+K$. One advantage of the projection operator approach is that it is not necessary to find an operator that annihilates S; nor is it necessary to filter out extra functions as the operator approach sometimes requires. In the operator approach a set of functions, the null set of the operator is annihilated. This set may contain functions other than those that one wishes to annihilate, that is, extra functions.

In these two approaches the practice is as follows: carry out PCA on the filtered data matrix XF based on the usual metric (I) or equivalently carry out PCA on the data matrix X based on the metric $M=FF'$. Generalized PCA with respect to a metric M may be stated as follows: given X $(n\times p)$ of full rank with $(n\geq p)$ find A and B which minimize $\|X-AB'\|$ subject to (i) A is $n\times r$, B is $p\times r$, with $r\leq p$; (ii) B is M-orthogonal (that is $B'MB=I$); (iii) the columns of A are in decreasing order. If M is singular, the solution is not unique. To find A and B', let $Y=XE\Lambda$, where $M=E\Lambda^2E'$. Take the SVD of $Y=PDU'$. Then $A=PD$ and $B'=Q'\Lambda^{-1}E'$.

In the projection operator approach the annihilation procedure leaves precisely the residue. Res. that is the difference between X and its projection on S. However in the differential and difference operator approach it leaves not Res. but the image of Res. under F.

3.2. Example

Since precisely Res. is obtained from only the projection operator approach, this approach was selected to demonstrate how filtering works in practice using the temperature data mentioned above. As can be seen by looking at Figure 6.7, once the obvious constant and cosinusoidal components are removed, three interesting phenomena become evident. The first eigenfunction corresponds to a warm spring, and is characteristic of the inland cities as opposed to coastal cities as can be seen from Figure 6.8. The second eigenfunction corresponds to an early spring, and is characteristic of cities which lie in a narrow latitude band namely

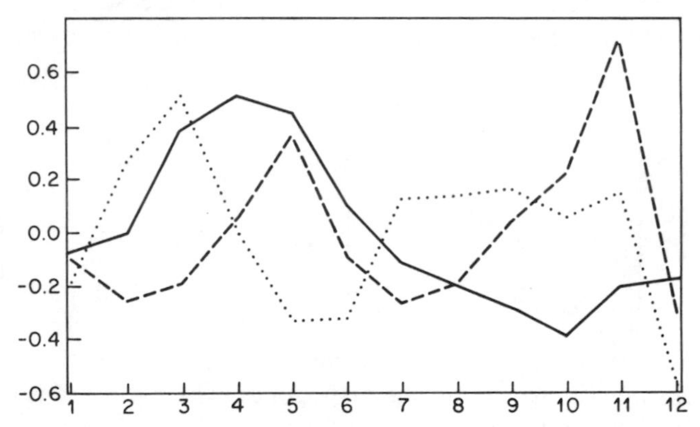

Figure 6.7. Plot of the first three principal components resulting from a PCA of the monthly temperature data based on the metric $M=FF'$, where $F=I-K^{\circ}K$ and $S=\{1, \cos\}$

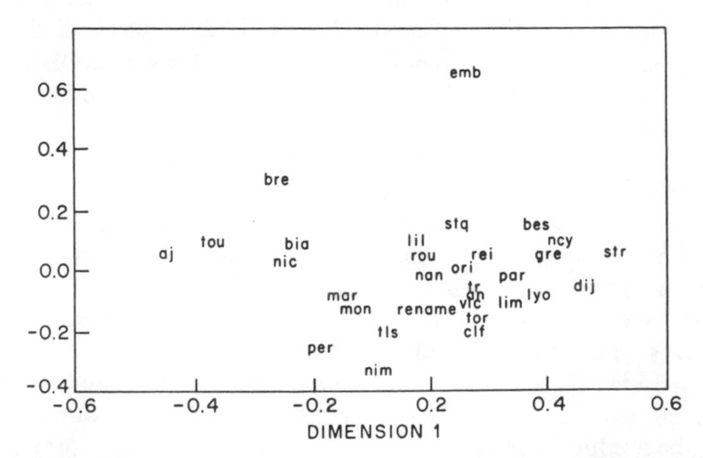

Figure 6.8. Biplot of the scores of the first and third principal component resulting from a PCA of the monthly temperature data based on a metric $M=FF'$, where $F=I-K^{\circ}K$ and $S=\{1, \cos\}$. The symbols correspond to the first three letters of the French city

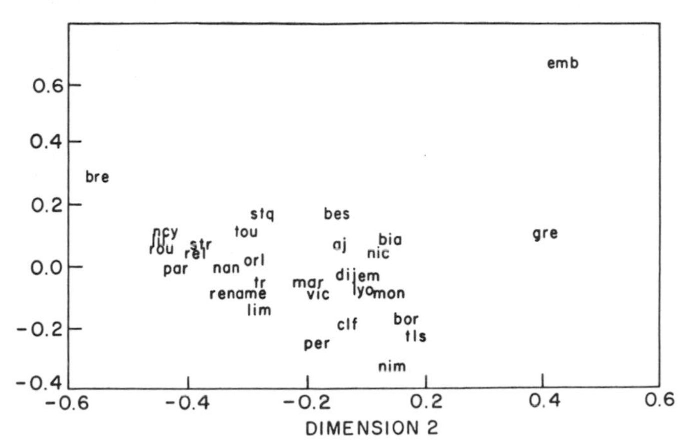

Figure 6.9. Biplot of the scores of the second and third principal component resulting from a PCA of the monthly temperature data based on a metric $M = FF'$, where $F = I - K^{\cup}K$ and $S = \{1, \cos\}$. The symbols correspond to the first three letters of the French city

Embrun, Grenoble, Toulouse, Bordeaux and Biarritz, as opposed to another narrow latitude band further north containing Brest, Nancy, Lille, Rouen and Paris (see Figure 6.8). The third eigenfunction corresponds to a sharp dip in temperature from November to December. This dip separates Embrun the only city in the sample at a higher altitude from the rest (see Figure 6.9). It appears that filtering is useful when trying to uncover subtle variation in a set of data consisting of sampled smooth functions.

References

Agarwal, G.G. and Studden, W.J. (1978). Asymptotic design and estimation using linear splines, *Commun. Statist.-Simula. Computa B*7, **4,** 309–319.

Andrews, D.F., Gnanadesikan, R. and Warner, J.L. (1971). Transformation of multivariate data, *Biometrika*, **27,** 825–40.

Aronszajn, N. (1950). Theory of reproducing kernel, *Transactions of the American Mathematical Society*, **68,** 337–404.

Bartlett, M.S. (1953). *Factor Analysis in Psychology as a Statistician Sees it.* Uppsala Symposium on Psychol. Factor Analysis. Uppsala: Almqvist & Wicksell.

Bekker, P. (1983). *Relationships Between Versions of Nonlinear Principal Component Analysis*. Leiden: Department of Data Theory.

Besse, P. (1979). *Etude descriptive des processus: Approximation et interpolation*. Thèse de 3ème Cycle. Toulouse: Université Paul Sabatier.

Besse, P. (1980). Deux exemples d'Analyse en Composantes Principales filtrantes, *Statistique et Analyse des Données*, **3,** 5–15.

Besse, P. (1986). Choice of a metric or choice of a model. The case of discrete events data. In F. de Antoni *et al.* (eds), *COMPSTAT. Proceedings in Computational Statistics.* Vienna: Physica-Verlag.

Besse, P. (1987). Choix de la métrique pour l'a.c.p. de séries d'évènements discrets, *Statistique et Analyse des données*, to appear.

Besse, P., Caussinus, H., Ferre, L. and Fine, J. (1987). Principal components analysis and optimization of graphical displays, submitted to *Statistics*.

Besse, P. and Ramsay, J.O. (1986). Principal component analysis of sampled functions, *Psychometrika*, **51,** 2, 285–311.

Besse, P. and Vidal, C. (1982). Analyse des correspondances et codage par une Probabilité de transition, *Statistique et Analyse des Données*, **7,** 3.

Benzécri, J.P. *et al.* (1973). *Analyse des données* (2 vols). Paris: Dunod.

Benzécri, J.P. (1980). *Pratique de l'analyse des données* 1. Paris: Dunod.

Bezdek, J.C. (1987). Some nonstandard clustering algorithms. In P. Legendre and L. Legendre (eds), *Developments in Numerical Ecology*. New York: Springer.

Boneva, L., Kendall, D. and Stefanov, I. (1971). Spline transformations: three new diagnostic aids for the data analyst, *JRSS Ser B.*, **33,** 1–70.

Bordet, J.P. (1973). *Etudes de données geophysiques*. Thèse de 3ème Cycle. Paris: Université de Paris VI.

Bowman, A.W. (1980). A note on consistency of the kernel method for the analysis of categorical data. *Biometrika*, **67,** 3.

Box, G.E.P. and Cox, D.R. (1964). Analysis of transformations (with discussion), *Journal of the Royal Statistical Society, Series B*, **26,** 211–52.

Breiman, L. and Friedman, J.H. (1985). Estimating optimal transformations for multiple regression and correlation. *JASA*, **80,** 580–598.

Burt, C. (1950). The factorial analysis of qualitative data, *British Journal of Statistical Psychology*, **3,** 166–85.

Carroll, J. D. (1968). A generalization of canonical correlation analysis to three or more sets of variables, *Proc. 76th Conv. APA*, 227–8.

Cattell, R. B. (1966). The scree test for the number of factors. *J. Multiv. Behav. Res.*, **1,** 245–76.

Chang, J.C. and Bargmann, R.E. (1974). *Internal Multidimensional Scaling of Categorical Variables*. Department of Statistics and Computer Science. Technical Report, University of Georgia.

Cliff, N. (1966). Orthogonal rotation to congruence, *Psychometrika*, **31,** 33–42.

Cogburn, R. and Davis, H.T. (1974). Periodical splines and spectral estimation, *Annals of Statistics*, **2,** 1108–26.

Conte, S. and De Boor, C. (1980). *Elementary Numerical Analysis*. 2nd edition. New York: McGraw-Hill.

Coolen, H., Van Rijckevorsel, J. and De Leeuw, J. (1982). An algorithm for nonlinear principal components analysis with B-splines by means of alternating least squares. In H. Caussinus *et al.* (eds), *COMPSTAT* 1982, part II. Vienna: Physika-Verlag.

Craddock, J.M. and Flood, C.R. (1969). Eigenvectors for representing the 500 mb. geopotential surface over the Northern Hemisphere. *Q.J.R. Met. Soc.*, **95,** 576–93.

Daudin, J.J., Duby, C. and Trecourt, P. (1987). Stability of principal components analysis studied by the bootstrap method, *Statistics*, to appear.

Dauxois, J. and Pousse, A. (1976). *Les analyses factorielles en calcul de probabilités et en statistique: Essai d'étude synthetique*. Thèse d'Etat. Toulouse: Université de Toulouse III.

De Boor, C. (1978). *A Practical Guide to Splines*. New York: Springer.

De Leeuw, J. (1973). *Canonical Analysis of Categorical Data*. Leiden: Psychological Institute. Reissued by DSWO Press, Leiden, 1984.

De Leeuw, J. (1976). *HOMALS*. Paper presented at the Psychometric Society meeting, Murray Hill, NJ.

De Leeuw, J. (1982). Nonlinear principal component analysis. In H. Caussinus *et al.* (eds), *COMPSTAT* 82. *Proceedings in Computational Statistics*, part I. Vienna: Physika-Verlag.

De Leeuw, J. (1983). On the prehistory of correspondence analysis, *Statistica Neerlandica*, **37,** 161–164.

De Leeuw, J. (1984a). The Gifi-system of non-linear multivariate analysis. In E. Diday, M. Jambu, L. Lebart, J. Pages and R. Tomassone (eds), *Data Analysis and Informatics*, volume III. Amsterdam: North-Holland.

De Leeuw, J. (1984b). Statistical properties of multiple correspondence analysis. Submitted for publication.

De Leeuw, J. (1984c). Models of Data, *Kwantitatieve Methoden*, **5,** 17–31.

De Leeuw, J. and Tijssen, R. (1984). *Multivariate Analysis with Optimal Scaling*. Leiden: Department of Data Theory.

De Leeuw, J., Van Rijckevorsel, J. and Van der Wouden, H. (1981). Nonlinear principal component analysis with B-splines, *Methods of Operations Research*, **33,** 379–93.

De Leeuw, J. and Walter, J. (1977). *Optimal Scaling of Continuous Numerical Data*. Leiden: Department of Data Theory.

De Leeuw, J. and Van Rijckevorsel, J. (1980). Homals en Princals. In E. Diday *et al.* (eds), *Data Analysis and Informatics*. Amsterdam: North-Holland.

De Leeuw, J., Young, F.W. and Takane, Y. (1976). Additive structure in qualitative data: an alternating least squares method with optimal scaling features, *Psychometrika*, **41,** 471–503.

De Vore, R. (1977). Monotone approximation by splines, *SIAM J. Math. Anal.*, **8,** 891–905.

Deville, C. (1977). Un exemple catastrophique d'analyse factorielle et son explication. In E. Diday (ed), *Premières Journées Internationales, Analyse des Données et Informatique*. Versailles: Inria, pp. 557–65.

Duc-Jacquet, M. (1973). *Approximation des fonctionnelles linéaires sur des espaces hilbertiens auto-reproduisants*, unpublished doctoral dissertation, Grenoble.

Eagle, A. (1928). On the relation between the fourier constants of a periodic function and the coefficients determined by harmonic analysis. *Phil. Mag.*, **5,** 113–32.

Eckart, C. and Young, G. (1936). The approximation of one matrix by another of lower rank. *Psychometrika*, **1,** 211–18.

Fischer, G.H. (1974). *Einführung in die Theorie psychologischer Tests, Grundlagen und Anwendungen*. Berne: Verlag Hans Huber.

Gallego, F.J. (1980). *Un codage flou pour l'analyse des correspondances*. Thèse de 3ème Cycle. Paris: Université de Paris VI.

Gautier, J.M. and Saporta, G. (1982). About fuzzy discrimination. In H. Caussinus *et al.* (eds), *Compstat* 1982. *Proceedings in Computational Statistics*. Vienna: Physica-Verlag.

Ghermani, B.M., Roux, C. and Roux, M. (1977). Sur le codage logique des données héterogènes, *Les Cahiers de l'Analyse des Données*, **1,** 115–18.

Gifi, A. (1981a). *Non-linear Multivariate Analysis*. Leiden: Department of Data Theory.

Gifi, A. (1981b). *HOMALS User's Guide*. Leiden: Department of Data Theory.

Gifi, A. (1982). *PRINCALS User's Guide*. Leiden: Department of Data Theory.

Gifi, A. (1988). *Non-linear Multivariate Analysis*. Leiden: DSWO Press.

Good, I.J. and Gaskins, R.A. (1971). Non-parametric roughness penalties for probability densities, *Biometrika*, **58,** 255–77.

Greenacre, M.J. (1981). Practical correspondence analysis. In V. Barnett (ed), *Interpreting Multivariate Data*. New York: Wiley.

Greenacre, M.J. (1984). *Theory and Applications of Correspondence Analysis*. New York: Academic Press.

Guitonneau, G.G. and Roux, M. (1977). Sur la taxinomie du genre erodium, *Les Cahiers de l'Analyse des Données*, **1,** 97–113.

Guttman, L. (1941). The quantification of a class of attributes: A theory and method of scale construction. In P. Horst *et al.* (eds), *The Prediction of Personal Adjustment*. New York: Social Science Research Council.

Harman, H.H. (1976). *Modern Factor Analysis*. Chicago: University of Chicago Press (revised edn).

Hayashi, C. (1952). On the prediction of phenomena from qualitative data and the quantification of qualitative data from the mathematico-statistical point of view. *Ann. Inst. Statist. Math.*, **5,** 121–143.

Holmquist, N.D., McMahan, C.A. and Williams, O.D. (1967). Variability in classification of carcinoma in situ of the uterine cervix, *Arch. Path.*, **84,** 334–45.

Horst, P. (1965). *Factor Analysis of Data Matrices*. New York: Holt, Rinehart & Winston.

Hudson, D.J. (1966). Fitting segmented curves whose join points have to be estimated, *JASA*, **61,** 1097–1129.

Jambu, M. and Lebeaux, M.O. (1983). *Cluster Analysis and Data Analysis*. Amsterdam: North-Holland.

Jolliffe, I. T. (1986). *Principal Component Analysis*. New York: Springer-Verlag.

Kettenring, J.R. (1971). Canonical analysis of several sets of variables, *Biometrika*, **58,** 433–60.

Koyak, R. (1985). *Optimal Transformations for Multivariate Linear Reduction Analysis*. Unpublished PhD thesis. University of California, Berkeley, California: Dept. of Statistics.

Kruskal, J.B. (1965). Analysis of factorial experiments by estimating monotone transformations of the data, *Journal of the Royal Statistical Society*, series B, **27**, 251–63.
Kruskal, J.B. and Shepard, R. N. (1974). A nonmetric variety of linear factor analysis, *Psychometrika*, **39**, 123–57.
Lafaye de Michaux, D. (1978). *Approximations d'analyses canoniques nonlineaires de variables aléatoires et analyses factorielles privelégiantes*. Thèse de Docteur-Ingenieur. Nice: Université de Nice.
Lancaster, H.O. (1969). *The Chi-squared Distribution*. New York: Wiley.
Lebart, L., Morineau, A. and Warwick, K.M. (1984). *Multivariate Descriptive Statistical Analysis*. New York: Wiley.
Le Foll, Y. (1979). *Sur les propriétés de l'analyse des correspondances pour diverses formes complètes de données*. Thèse de 3ème Cycle. Paris: Université de Paris VI.
Mallet, J.L. (1982). Propositions for fuzzy characteristics functions in data analysis. In H. Caussinus *et al.* (eds), *Compstat* 82. *Proceedings in Computational Statistics*. Vienna: Physica-Verlag.
Martin, J.F. (1980) *Le codage flou et ses applications en statistique*. Thèse de 3ème Cycle. Pau: Université de Pau et des Pays de l'Adour.
Martin, J.F. (1981). Formalisation et étude de codages flous, *C.R. Acad. Sci. Paris*, **292**, 299–301.
Maung, K. (1941). Measurement of association in a contingency table with special reference to the pigmentation of hair and eye colors of Scottish school children. *Ann. Eugen.*, **11**, 189–223.
McDonald, R.P. (1968). A unified treatment of the weighting problem, *Psychometrika*, **33**, 351–81.
Munoz-Perez, F. (1982). L'évolution de la fécondité dans les pays industrialisés dépuis 1971, *Population*, 37iéme année, **3**, 483–512.
Nishisato, S. (1980). *Analysis of Categorical Data: Dual Scaling and its Applications*. Toronto: University of Toronto Press.
Park, S.H. (1978). Experimental designs for fitting segmented polynomial regression models, *Technometrics*, **20**, 2, 151–4.
Poirier, D.J. (1973). Piecewise regression using cubic splines. *JASA*, **68**, 515–24.
Poirier, D.J. (1976). *The Econometrics of Structural Change*. Amsterdam: North-Holland.
Quade, W. and Collatz, L. (1938). Zur Interpolationstheorie der Reelen Periodischen Funktionen, *Aka, Wiss.* (*Math-Phys. Kl*), **30**, 383–429.
Rao, C.R. (1964). The use and interpretation of Principal Component Analysis in applied research, *Sankhyā*, **26(A)**, 329–58.
Ramsay, J.O. (1977). Monotonic weighted power transformations to additivity, *Psychometrika*, **42**, 83–109(a).
Ramsay, J.O. (1982). When data are functions, *Psychometrika*, **47**, 379–96.
Runge, C. (1901). Uber Die Darstellung Willkürlichen Funktionen und die Interpolation zwischen Aquidistanten Ordinaten, *Z. Angew, Math. Phys.*, **46**, 224–43.
Ruspini, E. (1969). A new approach to clustering, *Inf. and Control*, **15**, 22–32.
Saporta, G. (1985). Data analysis for numerical and categorical individual time-series. *Applied Stochastic Models and Data Analysis*, **1**, 109–119.
Schoenberg, I.J. (1946a). Contributions to the problem of approximation of equidistant data by analytic functions, Part A, *Quart. Appl. Math.*, **4**.
Schoenberg, I.J. (1946b). Contributions to the problem of approximation of equidistant data by analytic functions, Part B, *Quart. Appl. Math.*, **4**.
Schriever, B.F. (1986). *Order Dependence*, CWI-Tract 20. Amsterdam: Centre for Mathematics and Computer Science.
Schumaker, L. (1981). *Spline Functions: Basic Theory*. New York: Wiley.

Shapiro, H.S. (1971). *Topics in Approximation Theory*. New York: Springer-Verlag.

Smith, P. (1979). Splines as a useful statistical tool, *The American Statistician*, **33,** 57–62.

Styan, G.P. (1973). Hadamard products and multivariate statistical analysis, *Linear Algebra and its Applications*, **6,** 217–240.

Takane, Y., Young, F.W. and De Leeuw, J. (1978). The principal components of mixed measurement level multivariate data: an alternating least squares method with optimal scaling features, *Psychometrika*, **43,** 278–81.

Tenenhaus, M. (1977). Analyse en composantes principales d'un ensemble de variables nominales et numériques, *Revue de Statistique Appliqué*, **25,** 39–56.

Tenenhaus, M., and Young, F.W. (1985). An analysis and synthesis of multiple correspondence analysis, optimal scaling, dual scaling, homogeneity analysis and other methods for quantifying categorical multivariate data, *Psychometrika*, **50,** 91–119.

Thurstone, L.L. (1947). *Multiple Factor Analysis*. Chicago: University of Chicago.

Tijssen, R. (1984). *A New Approach to Nonlinear Canonical Correlation Analysis*. Leiden: Department of Data Theory.

Van de Geer, J.P. (1986). *Introduction to Linear Multivariate Analysis*. (2 Vols). Leiden: DSWO Press.

Van der Heijden, P.G.M. (1987). *Correspondence Analysis of Longitudinal Categorical Data*. Leiden: DSWO Press.

Van Rooij, P.L.J. and Schurer, F. (1974). A bibliography on spline functions. In K. Böhmer *et al.* (eds), *Tagung über Spline Funktionen*. Mannheim: Bigliographisches Institut, pp. 315–415.

van Rijckevorsel, J.L.A. (1982). Canonical analysis with B-splines. In H. Caussinus *et al.* (eds), *Compstat* 1982. *Proceedings in Computational statistics*. Vienna: Physica-Verlag, pp. 393–8.

van Rijckevorsel, J.L.A. (1987). *The Application of Horseshoes and Fuzzy Coding in Multiple Correspondence Analysis*. Leiden: DSWO Press.

van Rijckevorsel, J.L.A. and van Kooten, G. (1985). Smooth PCA of economic data, *Computational Statistics Quarterly*, **2,** 143–72.

Wahba, G. (1971). A polynomial algorithm for density estimation, *Annals of Mathematical Statistics*, **42,** 1870–86.

Wahba, G. (1978). Improper priors, spline smoothing and the problem of guarding against model errors in regression, *JRSC*, Ser. B, **40,** 364–72.

Wegman, E.J. and Wright, I. (1983). Splines in statistics, *JASA*, **78,** 351–65.

Winsberg, S. and Kruskal, J.O. (1986). Easy to generate metrics for use with sampled functions. In F. De Antoni *et al.* (eds), *COMPSTAT* 86. Proceedings in Computational Statistics. Vienna: Physica-Verlag.

Winsberg, S. and Ramsay, J.O. (1980). Monotonic transformations to additivity using splines, *Biometrika*, **67,** 669–74.

Winsberg, S. and Ramsay, J.O. (1981). Analysis of pairwise preferences data using integrated B-splines, *Psychometrika*, **46,** 171–86.

Winsberg, S. and Ramsay, J.O. (1982). Monotone splines: a family of transformations useful for data analysis. In H. Caussinus *et al.* (eds), *Compstat* 82. *Proceedings in Computational Statistics*. Vienna: Physica Verlag, pp. 451–6.

Winsberg, S. and Ramsay, J.O. (1983). Monotone spline transformations for dimension reduction, *Psychometrika*, **48,** 575–95.

Wold, S. (1974). Spline functions in data analysis, *Technometrics*, **16,** 1–11.

Wold, H. and Lyttkens, E. (1969). Nonlinear iterative partial least squares (Nipals) estimation procedures, *Bulletin of the International Statistical Institute*, **43,** 29–51.

Wright, I.W. and Wegman, E.J. (1980). Isotonic, convex and related splines, *Annals of Statistics*, **8,** 1023–35.

Young, F.W., De Leeuw, J. and Takane, Y. (1976). Regression with qualitative variables: an alternating least squares method with optimal scaling features, *Psychometrika*, **41,** 505–529.

Young, F.W., Takane, Y. and De Leeuw, J. (1978). The principal components of mixed measurement level multivariate data: an alternating least squares methods with optimal scaling features, *Psychometrika*, **43,** 279–81.

Young, F.W. (1981). Quantitative analysis of qualitative data, *Psychometrika*, **46,** 357–88.

Young, F.W., De Leeuw, J. and Takane, Y. (1980). Quantifying qualitative data. In E. D. Lantermann and H. Feger (eds), *Similarity and Choice. Papers in Honour of Clyde Coombs*. Berne: Hans Huber.

Zadeh, L.A. (1965). Fuzzy sets, *Inf. and Control*, **8,** 338–53.

Index

Applied Probability and Statistics (Continued)

HOEL • Elementary Statistics, *Fourth Edition*
HOEL and JESSEN • Basic Statistics for Business and Economics, *Third Edition*
HOGG and KLUGMAN • Loss Distributions
HOLLANDER and WOLFE • Nonparametric Statistical Methods
IMAN and CONOVER • Modern Business Statistics
JESSEN • Statistical Survey Techniques
JOHNSON • Multivariate Statistical Simulation
JOHNSON and KOTZ • Distributions in Statistics
Discrete Distributions
Continuous Univariate Distributions—1
Continuous Univariate Distributions—2
Continuous Multivariate Distributions
JUDGE, HILL, GRIFFITHS, LÜTKEPOHL and LEE • Introduction to the Theory and Practice of Econometrics, *Second Edition*
JUDGE, GRIFFITHS, HILL, LÜTKEPOHL and LEE • The Theory and Practice of Econometrics, *Second Edition*
KALBFLEISCH and PRENTICE • The Statistical Analysis of Failure Time Data
KISH • Statistical Design for Research
KISH • Survey Sampling
KUH, NEESE, and HOLLINGER • Structural Sensitivity in Econometric Models
KEENEY and RAIFFA • Decisions with Multiple Objectives
LAWLESS • Statistical Models and Methods for Lifetime Data
LEAMER • Specification Searches: Ad Hoc Inference with Nonexperimental Data
LEBART, MORINEAU, and WARWICK • Multivariate Descriptive Statistical Analysis: Correspondence Analysis and Related Techniques for Large Matrices
LINHART and ZUCCHINI • Model Selection
LITTLE and RUBIN • Statistical Analysis with Missing Data
McNEIL • Interactive Data Analysis
MAGNUS and NEUDECKER • Matrix Differential Calculus with Applications
MAINDONALD • Statistical Computation
MALLOWS • Design, Data, and Analysis by Some Friends of Cuthbert Daniel
MANN, SCHAFER and SINGPURWALLA • Methods for Statistical Analysis of Reliability and Life Data
MARTZ and WALLER • Bayesian Reliability Analysis
MIKÉ and STANLEY • Statistics in Medical Research: Methods and Issues with Applications in Cancer Research
MILLER • Beyond ANOVA, Basics of Applied Statistics
MILLER • Survival Analysis
MILLER, EFRON, BROWN, and MOSES • Biostatistics Casebook
MONTGOMERY and PECK • Introduction to Linear Regression Analysis
NELSON • Applied Life Data Analysis
OSBORNE • Finite Algorithms in Optimization and Data Analysis
OTNES and ENOCHSON • Applied Time Series Analysis: Volume I, Basic Techniques
OTNES and ENOCHSON • Digital Time Series Analysis
PANKRATZ • Forecasting with Univariate Box-Jenkins Models: Concepts and Cases
PLATEK, RAO, SARNDAL and SINGH • Small Area Statistics: An International Symposium
POLLOCK • The Algebra of Econometrics
RAO and MITRA • Generalized Inverse of Matrices and Its Applications
RÉNYI • A Diary on Information Theory
RIPLEY • Spatial Statistics

UNIVERSITY OF ROCHESTER LIBRARIES
3 9087 01037466 0